Kerry Young, Dan Evans and Ron Holt

ESSENTIALS

AQA GCSE

Additional Science

Contents

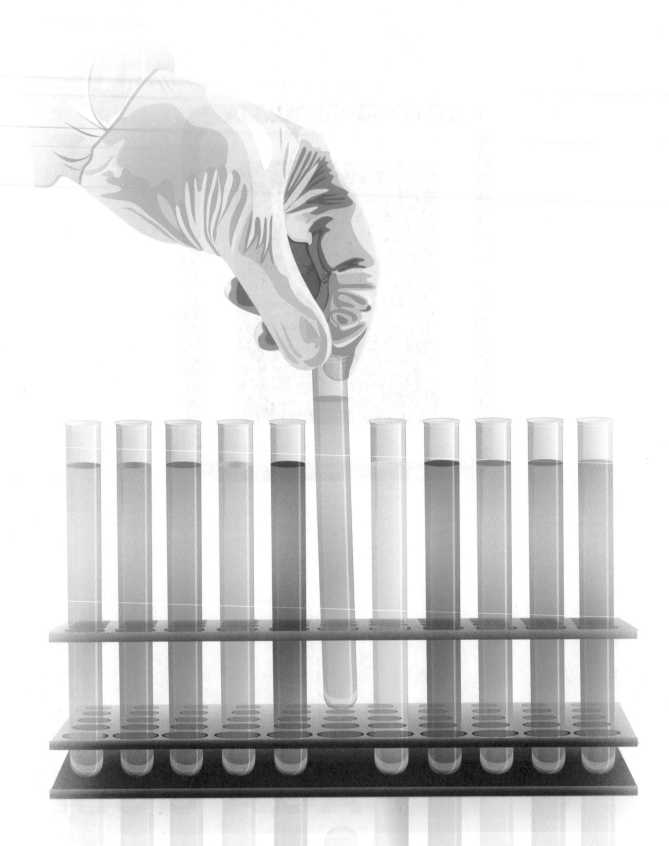

Contents

N.B. The numbers in brackets correspond to the reference numbers on the AQA GCSE Additional Science specification.

How Science Works – Explanation

The AQA GCSE Additional Science specification incorporates:

- **Science Content** – all the scientific explanations and evidence that you need to know for the exams. (It is covered on pages 12–79 of this revision guide.)
- **How Science Works** – a set of key concepts, relevant to all areas of science. It covers...
  - the relationship between scientific evidence, and scientific explanations and theories
  - how scientific evidence is collected
  - how reliable and valid scientific evidence is
  - the role of science in society
  - the impact science has on our lives
  - how decisions are made about the ways science and technology are used in different situations, and the factors affecting these decisions.

Your teacher(s) will have taught these two types of content together in your science lessons. Likewise, the questions on your exam papers will probably combine elements from both types of content. So, to answer them, you'll need to recall and apply the relevant scientific facts and knowledge of how science works.

The key concepts of How Science Works are summarised in this section of the revision guide (pages 5–11). You should be familiar with all of these concepts. If there is anything you are unsure about, ask your teacher to explain it to you.

How Science Works is designed to help you learn about and understand the practical side of science. It aims to help you develop your skills when it comes to...

- evaluating information
- developing arguments
- drawing conclusions.

The Thinking Behind Science

Science attempts to explain the world we live in.

Scientists carry out investigations and collect evidence in order to...

- **explain phenomena** (i.e. how and why things happen)
- **solve problems** using evidence.

Scientific knowledge and understanding can lead to the **development of new technologies** (e.g. in medicine and industry), which have a huge impact on **society** and the **environment**.

The Purpose of Evidence

Scientific evidence provides **facts** that help to answer a specific question and either **support** or **disprove** an idea or theory. Evidence is often based on data that has been collected through **observations** and **measurements**.

To allow scientists to reach conclusions, evidence must be...

- **repeatable** – other people should be able to repeat the same process
- **reproducible** – other people should be able to reproduce the same results
- **valid** – it must be repeatable, reproducible and answer the question.

N.B. If data isn't repeatable and reproducible, it can't be valid.

To ensure scientific evidence is repeatable, reproducible and valid, scientists look at ideas relating to...

- observations
- investigations
- measurements
- data presentation
- conclusions and evaluation.

How Science Works Overview

Observations

Most scientific investigations begin with an **observation**. A scientist observes an event or phenomenon and decides to find out more about how and why it happens.

The first step is to develop a **hypothesis**, which suggests an explanation for the phenomenon. Hypotheses normally suggest a relationship between two or more **variables** (factors that change).

Hypotheses are based on…
- careful observations
- existing scientific knowledge
- some creative thinking.

The hypothesis is used to make a **prediction**, which can be tested through scientific investigation. The data collected from the investigation will…
- support the hypothesis **or**
- show it to be untrue (refute it) **or**
- lead to the modification of the original hypothesis or the development of a new hypothesis.

If the hypothesis and models we have available to us do not completely match our data or observations, we need to check the validity of our observations or data, or amend the models.

Sometimes, if the new observations and data are valid, existing theories and explanations have to be revised or amended, and so scientific knowledge grows and develops.

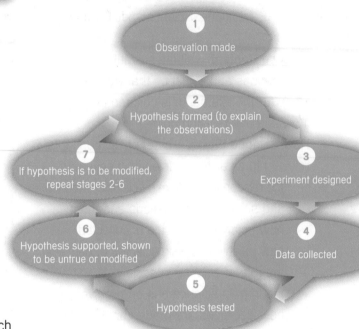

1. Observation made
2. Hypothesis formed (to explain the observations)
3. Experiment designed
4. Data collected
5. Hypothesis tested
6. Hypothesis supported, shown to be untrue or modified
7. If hypothesis is to be modified, repeat stages 2-6

Example

- Two scientists **observe** that freshwater shrimp are only found in certain parts of a stream.
- They use scientific knowledge of shrimp and water flow to develop a **hypothesis**, which relates the presence of shrimp (dependent variable) to the rate of water flow (independent variable). For example, a hypothesis could be: the faster the water flows, the fewer shrimp are found.
- They **predict** that shrimp are only found in parts of the stream where the water flow rate is below a certain value.
- They **investigate** by counting and recording the number of shrimp in different parts of the stream, where water flow rates differ.
- The **data** shows that more shrimp are present in parts of the stream where the flow rate is below a certain value. So, the data **supports** the hypothesis. But, it also shows that shrimp aren't always present in these parts of the stream.
- The scientists realise there must be another factor affecting the distribution of shrimp. They **refine their hypothesis**.

Investigations

An **investigation** involves collecting data to find out whether there is a relationship between two **variables**. A variable is a factor that can take different values.

In an investigation there are two types of variables:
- **Independent** variable – can be changed by the person carrying out the investigation. For example, the amount of water a plant receives.
- **Dependent** variable – measured each time a change is made to the independent variable, to see if it also changes. For example, the growth of the plant (measured by recording the number of leaves).

For a measurement to be valid it must measure only the appropriate variable.

Variables can have different types of values:
- **Continuous variables** – can take any numerical value (including decimals). These are usually measurements, e.g. temperature.
- **Categoric variables** – a variable described by a label, usually a word, e.g. different breeds of dog or blood group.
  - **Discrete variables** – only take whole-number values. These are usually quantities, e.g. the number of shrimp in a stream.
  - **Ordered variables** – have relative values, e.g. 'small', 'medium' or 'large'.

N.B. Numerical values, such as continuous variables, tend to be more informative than ordered and categoric variables.

An investigation tries to find out whether an **observed** link between two variables is…
- **causal** – a change in one variable causes a change in the other, e.g. the more cigarettes you smoke, the greater the chance that you will develop lung cancer.
- **due to association** – the changes in the two variables are linked by a third variable, e.g. as grassland decreases, the number of predators decreases (caused by a third variable, i.e. the number of prey decreasing).
- **due to chance** – the change in the two variables is unrelated; it is coincidental, e.g. people who eat more cheese than others watch more television.

Controlling Variables

In a **fair test**, the only factor that should affect the dependent variable is the independent variable. Other **outside variables** that could influence the results are kept the same, i.e. constant (control variables) or eliminated.

It's a lot easier to control all the other variables in a laboratory than in the field, where conditions can't always be controlled. The impact of an outside variable (e.g. light intensity or rainfall) has to be reduced by ensuring all the measurements are affected by it in the same way. For example, all the measurements should be taken at the same time of day.

Control groups are often used in biological and medical research to make sure that any observed results are due to changes in the independent variable only.

A sample is chosen that 'matches' the test group as closely as possible except for the variable that is being investigated, e.g. testing the effect of a drug on reducing blood pressure. The control group must be the same age, gender, have similar diets, lifestyles, blood pressure, general health, etc.

Investigations (Cont.)

Accuracy and Precision

How accurate data needs to be depends on what the investigation is trying to find out. For example, when measuring the volume of acid needed to neutralise an alkaline solution it is important that equipment is used that is able to accurately measure volumes of liquids.

The data collected must be **precise** enough to form a **valid conclusion**: it should provide clear evidence for or against the hypothesis.

Measurements

Apart from control variables, there are a number of factors that can affect the reliability and validity of measurements:

- **Accuracy of instruments** – depends on how accurately the instrument has been calibrated. An accurate measurement is one that is close to the true value.
- **Resolution (or sensitivity) of instruments** – determined by the smallest change in value that the instrument can detect. The more sensitive the instrument, the more **precise** the value. For example, bathroom scales aren't sensitive enough to detect changes in a baby's mass, but the scales used by a midwife are.
- **Human error** – even if an instrument is used correctly, human error can produce random differences in repeated readings or a systematic shift from the true value if you lose concentration or make the same mistake repeatedly.
- **Systematic error** – can result from repeatedly carrying out the process incorrectly, making the same mistake each time.
- **Random error** – can result from carrying out a process incorrectly on odd occasions or by fluctuations in a reading. The smaller the random error the greater the accuracy of the reading.

To ensure data is as accurate as possible, you can…

- calculate the **mean** (average) of a set of repeated measurements to reduce the effect of random errors
- increase the number of measurements taken to improve the reliability of the mean / spot anomalies.

Preliminary Investigations

A trial run of an investigation will help identify appropriate values to be recorded, such as the number of repeated readings needed and their range and interval.

You need to examine any **anomalous** (irregular) values to try to determine why they appear. If they have been caused by equipment failure or human error, it is common practice to ignore them and not use them in any calculations.

There will always be some variation in the actual value of a variable, no matter how hard we try to repeat an event.

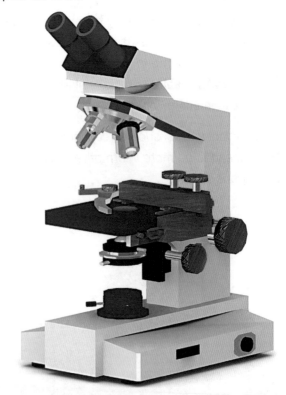

Presenting Data

Data is often presented in a **chart** or **graph** because it makes...

- any patterns more obvious
- it easier to see the relationship between two variables.

The **mean** (or average) of data is calculated by adding all the measurements together, then dividing by the number of measurements taken:

$$\text{Mean} = \frac{\text{Sum of all Values}}{\text{Number of Values}}$$

If you present data clearly, it is easier to identify any anomalous (irregular) values. The type of chart or graph you use to present data depends on the type of variable involved:

1 Tables organise data (but patterns and anomalies aren't always obvious)

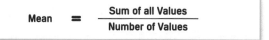

Height of student (cm)	127	165	149	147	155	161	154	138	145
Shoe size	5	8	5	6	5	5	6	4	5

2 Bar charts display data when the independent variable is categoric or discrete and the dependent variable is continuous.

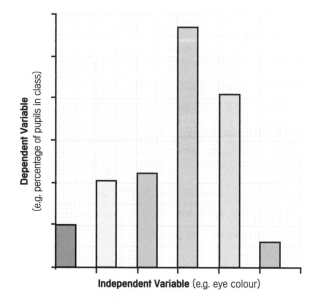

Dependent Variable (e.g. percentage of pupils in class)

Independent Variable (e.g. eye colour)

3 Line graphs display data when both variables are continuous.

- Points are joined by straight lines if you don't have data to support the values between the points.
- A line of best fit is drawn if there is sufficient data or if a trend can be assumed.

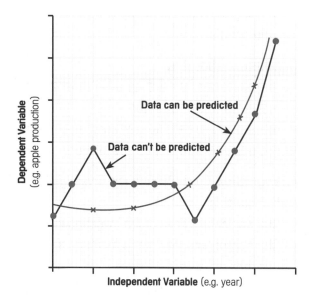

Dependent Variable (e.g. apple production)

Data can be predicted

Data can't be predicted

Independent Variable (e.g. year)

4 Scattergrams (scatter diagrams) show the underlying relationship between two variables. This can be made clearer if you include a **line of best fit**. A line of best fit could be a straight line or a smooth curve.

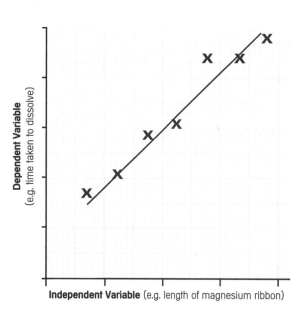

Dependent Variable (e.g. time taken to dissolve)

Independent Variable (e.g. length of magnesium ribbon)

Conclusions and Evaluations

Conclusions **should**…
- describe patterns and relationships between variables
- take all the data into account
- make direct reference to the original hypothesis or prediction
- try to explain the results / observations by making reference to the hypothesis as appropriate.

Conclusions **should not**…
- be influenced by anything other than the data collected (i.e. be biased)
- disregard any data (except anomalous values)
- include any unreasoned speculation.

An **evaluation** looks at the whole investigation. It should consider…
- the original purpose of the investigation
- the appropriateness of the methods and techniques used
- the reliability and validity of the data
- the validity of the conclusions.

The **reliability** of an investigation can be increased by…
- looking at relevant data from secondary sources (i.e. sources created by someone who did not experience first hand or participate in the original experiment)
- using an alternative method to check results
- ensuring results can be reproduced by others.

Science and Society

Scientific understanding can lead to technological developments. These developments can be exploited by different groups of people for different reasons. For example, the successful development of a new drug…
- benefits the drugs company financially
- improves the quality of life for patients
- can benefit society (e.g. if a new drug works, then maybe fewer people will be in hospital, which reduces time off sick, cost to the NHS, etc).

Scientific developments can raise certain **issues**. An issue is an important question that is in dispute and needs to be settled. The resolution of an issue may not be based on scientific evidence alone.

There are several different types of **issue** that can arise:
- **Social** – the impact on the human population of a community, city, country, or the world.
- **Economic** – money and related factors like employment and the distribution of resources.
- **Environmental** – the impact on the planet, its natural ecosystems and resources.
- **Ethical** – what is morally right or wrong; requires a value judgement to be made.

N.B. There is often an overlap between social and economic issues.

Peer Review

Finally, peer review is a process of self-regulation involving qualified professional individuals or experts in a particular field who examine the work undertaken critically. The vast majority of peer review methods are designed to maintain standards and provide credibility for the work that has been undertaken. These methods vary depending on the nature of the work and also on the overall purpose behind the review process.

Evaluating Information

It is important to be able to evaluate information relating to social-scientific issues, for both your GCSE course and to help you make informed decisions in life.

When evaluating information...
- make a list of **pluses** (pros)
- make a list of **minuses** (cons)
- consider how each point might **impact on society**.

You also need to consider whether the source of information is reliable and credible. Some important factors to consider are...
- **opinions**
- **bias**
- **weight of evidence**.

Opinions are personal viewpoints. Opinions backed up by valid and reliable evidence carry far more weight than those based on non-scientific ideas.

Opinions of experts can also carry more weight than non-experts.

Information is **biased** if it favours one particular viewpoint without providing a balanced account.

Biased information might include incomplete evidence or try to influence how you interpret the evidence.

Scientific evidence can be given **undue weight** or dismissed too quickly due to...
- political significance (consequences of the evidence could provoke public or political unrest)
- status of the experiment (e.g. if they do not have academic or professional status, experience, authority or reputation).

Limitations of Science

Although science can help us in lots of ways, it can't supply all the answers. We are still finding out about things and developing our scientific knowledge.

There are some questions that science can't answer. These tend to be questions...
- where beliefs, opinions and ethics are important
- where we don't have enough reproducible, repeatable or valid evidence.

Science can often tell us if something **can** be done, and **how** it can be done, but it can't tell us whether it **should** be done.

Decisions are made by individuals and by society on issues relating to science and technology.

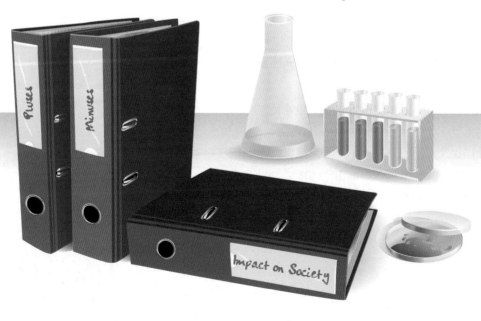

B2 Cells and Simple Cell Transport

Cells

All **living things** are made up of **cells**. The **structures** of different types of cells are related to their **functions**.

Cells may be **specialised** to carry out a particular job.

Root Hair Cells	Ovum (egg cell)	Xylem	White Blood Cells	Sperm Cells	Palisade Cells	Red Blood Cells	Nerve Cells
Tiny hair-like extensions that increase the surface area of the cell for absorption.	Large cell that can carry food reserves for the developing embryo.	Long, thin, hollow cells used to transport water through the stem and root.	Can change shape in order to engulf and destroy invading microorganisms.	Has a tail, which allows it to move.	Packed with chloroplasts for photosynthesis.	No nucleus, so packed full of haemoglobin to absorb oxygen.	Long, slender axons that can carry nerve impulses.

Cell Structure

Most human cells, animal cells and plant cells have the following parts:
- **Nucleus** – controls the activities of the cell.
- **Cytoplasm** – most chemical reactions take place here.
- A **cell membrane** – controls the passage of substances in and out of the cell.
- **Mitochondria** – where most energy is released in **respiration**.
- **Ribosomes** – where protein synthesis occurs.

Chemical reactions inside cells are controlled by **enzymes** found in the **cytoplasm** and **mitochondria**.

A Human Cheek Cell

Cell membrane
Mitochondria
Cytoplasm (contains mitochondria)
Nucleus
Ribosomes

Plant Cells

Plant cells also have the following parts:
- **A cell wall** – (made out of cellulose) used to strengthen the cell. (Algal cells also have a cell wall.)
- **Chloroplasts** – absorb light energy to make food.
- A **permanent** **vacuole** – filled with cell sap.

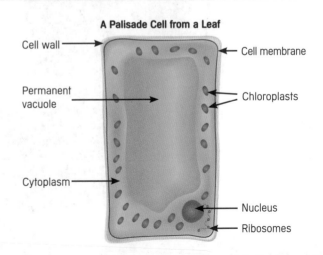

A Palisade Cell from a Leaf

Cell wall
Cell membrane
Permanent vacuole
Chloroplasts
Cytoplasm
Nucleus
Ribosomes

Key Words **Cell • Specialised • Nucleus • Cytoplasm • Mitochondria • Ribosome • Vacuole**

Other Cells

Singled celled organism.

In a bacterial cell, the genes are **not** in a nucleus.

A **bacterial** cell consists of...
- cytoplasm
- a cell membrane
- a cell wall.

Yeast is a single-celled fungus. Yeast cells have a...
- nucleus
- cytoplasm
- cell membrane
- cell wall.

A Single Bacterium

Cell wall → ← Cytoplasm
Genetic material → ← Cell membrane

Yeast

Vacuole → ← Nucleus
Cell wall → ← Membrane

Movement of Substances

Cells have to constantly...
- **replace** substances that are used up, e.g. glucose and **oxygen** for **respiration**
- **remove** other substances, e.g. carbon dioxide and waste products.

Gases and substances that are **in solution** (**dissolved**) can pass into and out of cells, through the cell membrane, by **diffusion**.

Oxygen
Glucose
Carbon dioxide
Waste products

Diffusion

Diffusion is the spreading of the **particles of a gas** or a substance in solution.

It results in a **net movement** from a region where the particles are of a **higher concentration** to a region where they are of a **lower concentration**.

The **greater the difference** in concentration, the **faster the rate** of diffusion.

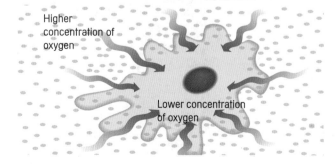

Higher concentration of oxygen

Lower concentration of oxygen

Quick Test

1. Which part of a cell controls the activities inside the cell?
2. Where are proteins made in a cell?
3. Where does photosynthesis take place in a plant cell?
4. Name the type of proteins that control chemical reactions inside the cell.
5. Name the process that takes place when oxygen moves from the lungs into the blood.
6. Name one substance that is removed from cells.

B2 Tissues, Organs and Organ Systems

Tissues

Large **multicellular** organisms, like humans, develop systems for **exchanging** materials. As the organism develops, cells **differentiate** so that they can carry out different jobs.

A **tissue** is a group of cells that have a **similar structure and function**. For example…

- **muscle tissue** contracts so we can move
- **glandular tissue** produce substances such as enzymes and hormones
- **epithelial tissue** covers organs.

Muscle Tissue
Can contract to bring about movement

Glandular Tissue
Can produce substances such as enzymes and hormones

Epithelial Tissue
Covers all parts of the body

Organs

Organs are made of **tissues**. One organ may contain several tissues.

For example, the **stomach** is an organ that contains…

- **muscle tissue** that contracts to churn the contents
- **glandular tissue** to produce digestive juices
- **epithelial tissue** to cover the outside and inside of the stomach.

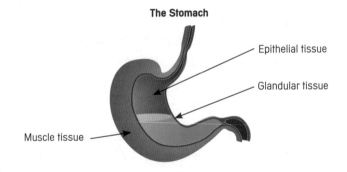

The Stomach

Epithelial tissue

Glandular tissue

Muscle tissue

Organ Systems

Organ systems are **groups** of organs that carry out a particular function.

For example, the **digestive system** includes…

- glands, such as the **pancreas** and **salivary glands**, that produce digestive juices
- the **stomach** and **small intestine**, where digestion takes place
- the **liver**, which produces **bile** to help break down fats
- the small intestine, where the **soluble food** is **absorbed** into the blood
- the large intestine, where **water is absorbed** from undigested food, producing faeces.

Humans also have other organ systems such as the excretory system involving the kidneys, and the reproductive system.

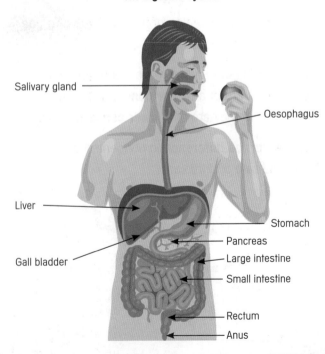

The Digestive System

Salivary gland

Oesophagus

Liver

Stomach

Pancreas

Large intestine

Small intestine

Gall bladder

Rectum

Anus

Key Words   **Multicellular • Differentiation • Tissue • Organ**

Plant Organs

Plant organs include...

- stems
- roots
- leaves.

Plant tissues include...

- **epidermal tissues**, that cover the plant
- **mesophyll**, where photosynthesis takes place
- **xylem and phloem** that transport substances around the plant.

Palisade cells contain chlorophyll for photosynthesis

Waxy cuticle – for waterproofing.

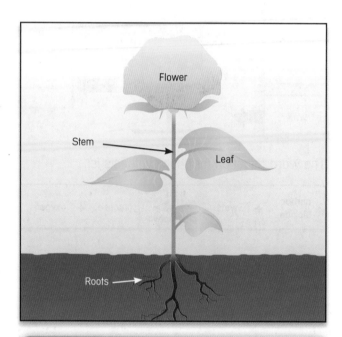

Flower
Stem
Leaf
Roots

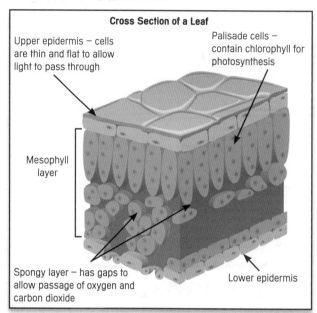

Cross Section of a Leaf

Upper epidermis – cells are thin and flat to allow light to pass through

Palisade cells – contain chlorophyll for photosynthesis

Mesophyll layer

Spongy layer – has gaps to allow passage of oxygen and carbon dioxide

Lower epidermis

Levels of Organisation

The following flow charts should help you to understand the scale and size of cells, tissues and organ systems.

Spongy layer has gaps to allow passage of oxygen + CO_2.

U + L Epidermis flat cells are thin to allow light to pass through.

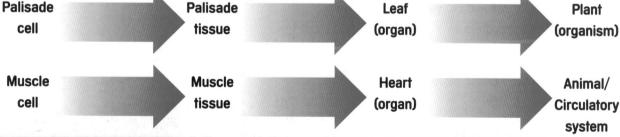

| Palisade cell | → | Palisade tissue | → | Leaf (organ) | → | Plant (organism) |

| Muscle cell | → | Muscle tissue | → | Heart (organ) | → | Animal/ Circulatory system |

Quick Test

1. What are a group of cells with a similar structure and function called?
2. Is the heart a tissue, organ or organ system?
3. In which plant organ would you expect to find photosynthesis taking place?
4. In which human organ system would you expect to find the brain?
5. Why do cells differentiate?

B2 Photosynthesis

Photosynthesis

Green plants don't absorb food from the soil. They make their own food using sunlight. This process is called **photosynthesis**.

Photosynthesis occurs in the cells of **green plants** and **algae** that are exposed to **light**.

During photosynthesis, light energy is absorbed by green **chlorophyll**, which is found in **chloroplasts** in some plant cells and algae.

Four things are needed for photosynthesis:
- **Light** from the Sun.
- **Carbon dioxide** from the air.
- **Water** from the soil.
- **Chlorophyll** in the leaves.

The **light energy** is used to convert **carbon dioxide** from the air and **water** from the soil into **sugar** (glucose). **Oxygen** is released as a by-product.

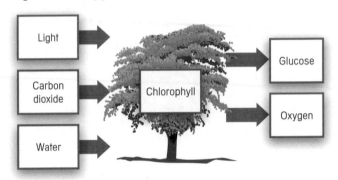

The word equation for photosynthesis is:

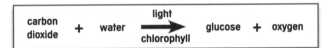

$$\text{carbon dioxide} + \text{water} \xrightarrow[\text{chlorophyll}]{\text{light}} \text{glucose} + \text{oxygen}$$

How Plants Use Glucose

The glucose produced in photosynthesis may be…
- changed into **insoluble starch** and stored in the stem, leaves or roots
- used by the plant during **respiration** to provide **energy**.

used to produce cellulose to strengthen cell walls.

Some glucose in plants and algae is…
- used to produce fat or oil and stored
- used to produce **cellulose** to **strengthen cell walls**
- used to produce **proteins**.

To produce proteins, plants also use **nitrate ions**, which are absorbed from the **soil**.

Factors Affecting Photosynthesis

There are several factors that may at any time **limit** the rate of photosynthesis:

- **Temperature**
- **Carbon dioxide** (CO_2) concentration
- **Light intensity**

Temperature

1. As the temperature increases, so does the rate of photosynthesis. Temperature limits the rate of photosynthesis.
2. As the temperature approaches 45°C, the rate of **photosynthesis** drops to zero because the **enzymes** controlling photosynthesis have been destroyed.

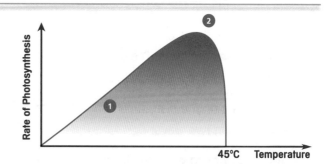

CO_2 Concentration

1. As the rate of carbon dioxide concentration increases, so does the rate of **photosynthesis**. CO_2 limits the rate of photosynthesis.
2. After reaching a certain point, an increase in CO_2 has no further effect. CO_2 is no longer the limiting factor – it must be either light or temperature.

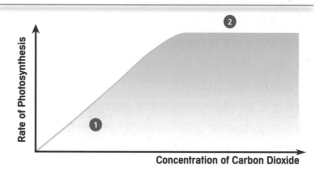

Light Intensity

1. As light intensity increases, so does the rate of photosynthesis. Light intensity limits the rate of photosynthesis.
2. After reaching a certain point, an increase in light intensity has no further effect. Light intensity is no longer the limiting factor – it must be either carbon dioxide or temperature.

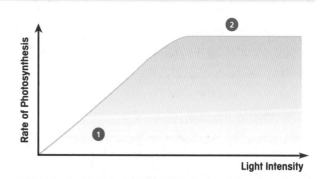

Artificial Controls

Greenhouses can be used to control the rate of photosynthesis. By controlling lighting, temperature and carbon dioxide, gardeners can increase the rate of photosynthesis.

This can result in plants...

- growing more quickly
- becoming bigger and stronger.

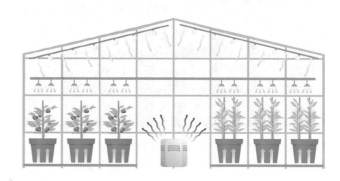

B2 Organisms and their Environment

Distribution of Organisms

There are a number of **physical** factors that affect organisms:
- **Temperature**
- Availability of **nutrients**
- Amount of **light**
- Availability of **water**
- Availability of **oxygen** and **carbon dioxide.**

Investigating the Distribution of Organisms

It is normally very difficult to count **all** the species in a habitat, so **samples** are taken. The size of the sample affects how valid and reproducible the data is.

Quantitative data on the distribution of organisms can be obtained in two ways.

Random Sampling with Quadrats

A student wants to know how many species of plants there are in her school field.
1. The student sets out a sample area, e.g. a school field (100m²).
2. A 1m² **quadrat** is placed randomly in the field.
3. The number of different plant species found in the quadrat is recorded.
4. This is **repeated** several times.
5. A **mean** is calculated from the data by adding up the number of species found in each quadrat and dividing by the number of quadrats used.
6. The total number of organisms in the school field can then be **estimated** by multiplying the mean number by the total area of the school field.

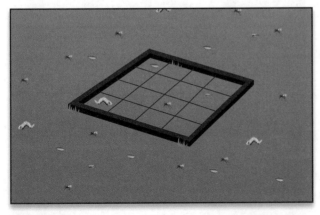

A quadrat can come in different shapes and sizes, but 1m² or 0.5m² are most commonly used.

Sampling Along a Transect

A student wants to find out if the shade of a big hedge in his school field affects the number of species of plants that are able to grow near it.
1. The student stretches some string from the hedge out into the field, producing a **transect**.
2. A quadrat is placed at 5m intervals along the transect moving away from the hedge.
3. The student records the number of plant species found in each quadrat along the transect.
4. To increase the **reliability** of the data, the student repeats his investigation by using more transects.

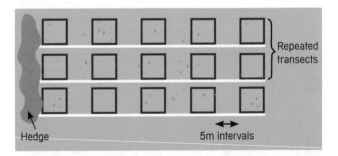

Repeated transects

Hedge

5m intervals

Quick Test

1. Name the green pigment essential for photosynthesis.
2. Where do plants obtain the carbon dioxide used in photosynthesis?
3. What do plants need, in addition to glucose, to make proteins?
4. List three factors that limit the rate of photosynthesis.
5. List two methods of sampling the distribution of daisies in a field.

Quantitative • Quadrat • Transect

Proteins

Protein molecules are made from long chains of **amino acids**. The chains are folded into a specific 3-D shape that allows other molecules to fit into the protein.

Proteins act as...

- **structural components** of tissues such as muscles
- **hormones**
- **antibodies**
- catalysts.

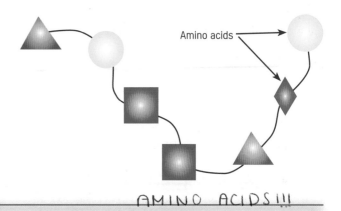

Protein Molecule Made up of Amino Acids

Amino acids

AMINO ACIDS!!!

Enzymes

Enzymes are **biological catalysts** made from proteins. Catalysts increase the rate of chemical reactions.

The **shape** of an enzyme is **vital** for its **function**. **High temperatures change** the shape of most enzymes. This is why it's dangerous for your body temperature to go much above 37°C.

Different enzymes work best at different temperatures and pH levels.

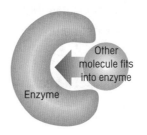

Enzymes

Other molecule fits into enzyme

Enzyme

Denatured

Enzyme destroyed by heat

Heat

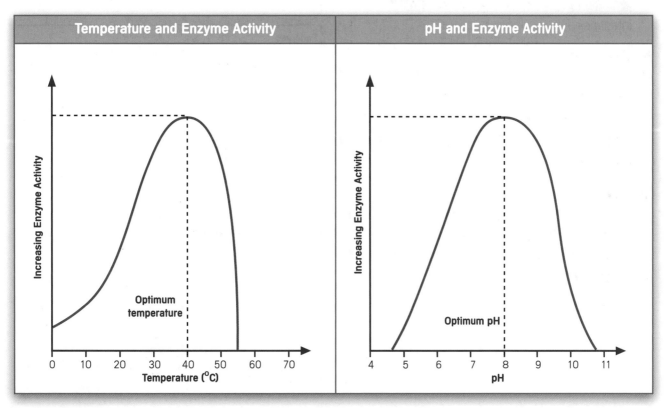

Temperature and Enzyme Activity	pH and Enzyme Activity

Increasing Enzyme Activity

Optimum temperature

Temperature (°C)

Increasing Enzyme Activity

Optimum pH

pH

B2 Proteins – their Functions and Uses

Enzymes Outside Body Cells

Some enzymes work **outside** body cells.

Digestive enzymes are produced by **specialised** cells in glands and in the lining of the gut.

The enzymes work as follows:

1. They pass out of the cells into the digestive system.
2. They come into contact with food molecules.
3. They catalyse the breakdown of large food molecules into smaller molecules.

The digestive enzymes **protease**, **lipase** and **amylase** are produced in four separate regions of the digestive system: **salivary glands**, **stomach**, **pancreas** and **small intestine**.

The enzymes **digest** proteins, fats and carbohydrates to produce **smaller molecules** that can be absorbed easily into the bloodstream.

Amylase…

- is produced in the salivary glands, pancreas and small intestine
- digests starch
- produces sugars in the mouth and small intestine.

Protease…

- is produced in the stomach, pancreas and small intestine
- digests proteins
- produces **amino acids** in the stomach and small intestine.

Lipase…

- is produced in the pancreas and small intestine
- digests lipids (fats and oils)
- produces fatty acids and glycerol in the small intestine.

Salivary Glands
Produces amylase, which starts to digest carbohydrates

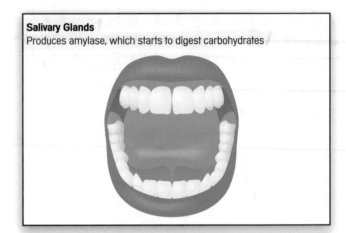

Stomach
Produces protease and also produces hydrochloric acid in which stomach enzymes work best

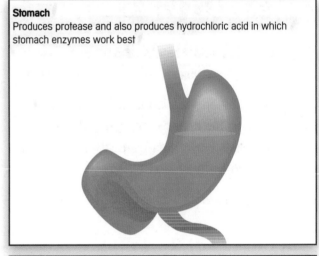

Pancreas
Produces all three types of enzyme that are then released into the small intestine

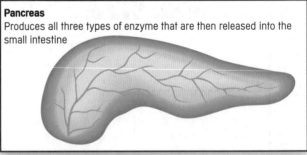

Small Intestine
Produces all three types of enzyme and absorbs small, soluble digested food molecules into the bloodstream

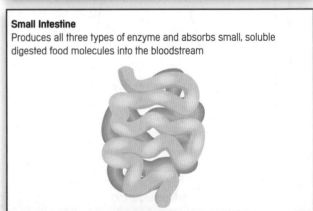

Key Words   **Protease • Lipase • Amylase • Amino acids**

Enzymes Outside Body Cells

Bile is **produced** in your **liver**. It is **stored** in your **gall bladder** before being released into the **small intestine**.

Bile **neutralises the acid** that is added to food in your stomach. This produces alkaline conditions in which **enzymes** in the small intestine work best.

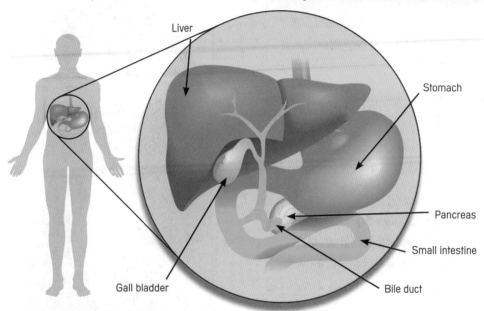

Liver

Stomach

Pancreas

Small intestine

Gall bladder

Bile duct

Uses of Enzymes

Some microorganisms produce enzymes that can be used to our benefit **in the home** and in **industry**.

In the home, **biological detergents**…
- may contain **protein-digesting** (protease) enzymes, to break down blood and food stains
- may contain **fat-digesting** (lipase) enzymes, to break down oil and grease stains
- are **more** effective at **lower** temperatures than other types of detergents.

In industry **various enzymes** are used, such as…
- **proteases**, to 'pre-digest' protein in some baby foods
- **carbohydrases**, to convert starch into sugar syrup
- **isomerase**, to convert glucose syrup into fructose syrup, which is sweeter and can be used in small quantities in slimming foods.

Enzymes are used in industry to bring about reactions at **normal temperatures** and **pressures**, which would otherwise need expensive, energy-demanding equipment. Most enzymes are **denatured** (lose their shape) at higher temperatures and many are expensive to produce.

Quick Test

1. What type of molecule is an enzyme?
2. List two factors that affect the rate at which enzymes work at.
3. Where are protease enzymes produced in the body?
4. What is produced when amylase digests starch?
5. What type of enzyme breaks down fats?
6. Where is bile produced and what is its main function?
7. Which two types of enzyme are found in biological detergents?

Aerobic Respiration

The chemical reactions inside cells are **controlled** by enzymes.

During **aerobic respiration** (respiration that uses oxygen) chemical reactions occur that...
- use **glucose** (a sugar) and **oxygen**
- release **energy**.

Aerobic respiration takes place continuously in both plants and animals.

Most of the reactions involved in aerobic respiration take place in the **mitochondria** in the cytoplasm of cells.

Energy that is released during respiration may be used...
- to **build larger molecules** from smaller ones
- to allow muscles to **contract** (in animals)
- to **maintain a steady body temperature** in colder surroundings (in mammals and birds)
- to build up amino acids from **sugars**, **nitrates** and other **nutrients**, which then combine to form proteins (in plants).

Aerobic respiration is summarised by the equation below:

Water lost in our breath is a waste product of aerobic respiration

glucose + oxygen ⟶	carbon dioxide +	water +	energy
Glucose and **oxygen** are brought to the respiring cells by the bloodstream.	**Carbon dioxide** is taken by the blood to the lungs and breathed out.	**Water** passes into the blood and is lost as sweat, moist breath and urine.	**Energy** is used for muscle contraction, metabolism and maintaining temperature.

A Working Muscle Cell

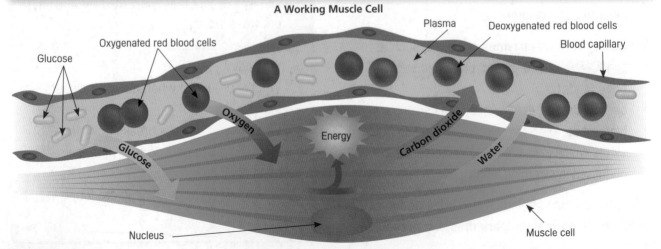

Glucose

Oxygenated red blood cells

Plasma

Deoxygenated red blood cells

Blood capillary

Oxygen

Glucose

Energy

Carbon dioxide

Water

Nucleus

Muscle cell

Exercise and the Body

During exercise a number of changes take place in your body:

- Your heart rate increases.
- The rate and depth of your breathing increases.
- The arteries supplying your muscles **dilate**.
- Bloodflow to your muscles increases.
- The supply of oxygen and sugar and the removal of carbon dioxide is increased.

Muscles store glucose as **glycogen**, which can then be converted back to glucose for use during exercise.

Anaerobic Respiration

Anaerobic respiration is summarised by the equation below:

glucose →	energy	+	lactic acid
Glucose from the bloodstream.	A **small** amount of energy is produced quickly and used for explosive activity.		**Lactic acid** accumulates in the muscles making them feel tired and 'rubbery'.

If there isn't enough oxygen reaching your muscles, they use **anaerobic respiration** to obtain energy.

Anaerobic respiration releases energy **without** oxygen. It is the **incomplete** breakdown of glucose and produces **lactic acid**.

If muscles carry out vigorous activity for a long time they become **fatigued** – they stop contracting efficiently and hurt.

Muscle fatigue is caused by the build up of lactic acid in the muscles. Blood flowing through the muscles will **remove** the lactic acid.

(HT) Anaerobic respiration releases much **less energy** than aerobic respiration (about one twentieth) because the breakdown of glucose is incomplete.

Lactic acid builds up in the muscles and must be oxidised into carbon dioxide and water. The oxygen needed to do this is called the oxygen debt.

Quick Test

1. In which part of the cell does respiration take place?
2. Which type of respiration requires oxygen?
3. What happens to the carbon dioxide produced during aerobic respiration?
4. Name the waste product of anaerobic respiration that accumulates in muscles.
5. (HT) What term describes the amount of oxygen 'owed' to the body after anaerobic respiration?

Key Words   Dilate • Glycogen • Anaerobic respiration • Lactic acid • Fatigued • Oxygen debt

B2 Cell Division and Inheritance

Human Body Cells

Body cells contain **two sets** of **23 chromosomes** arranged in pairs (46 in total).

Chromosomes contain genetic information.

Gametes are sex cells, i.e. female eggs and male sperm. Gametes only have **one set** of **23** chromosomes.

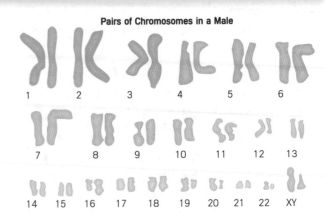

Pairs of Chromosomes in a Male

1 2 3 4 5 6
7 8 9 10 11 12 13
14 15 16 17 18 19 20 21 22 XY

Fertilisation

During **fertilisation**, the female and male gametes fuse to produce a **zygote** – a **single body cell** with **23 new pairs** of chromosomes. In each pair...

- one chromosome comes from the mother
- one chromosome comes from the father.

The cell then divides repeatedly by **mitosis** to form a new individual.

Variation is caused due to the combination of genes from the mother and father.

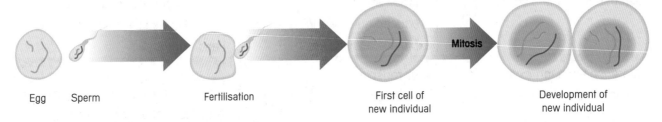

Egg   Sperm   Fertilisation

Mitosis

First cell of
new individual

Development of
new individual

Mitosis

Mitosis is the division of body cells to produce new cells. Mitosis occurs...

- for growth
- for repair
- in asexual reproduction (cells produced by asexual reproduction contain the same alleles as the parents).

During mitosis...

- a copy of each chromosome is made
- the cell then divides **once** to produce two body cells
- the **new cells** contain exactly the **same genetic information** as the **parent** cell, i.e. the same number of chromosomes and the same genes.

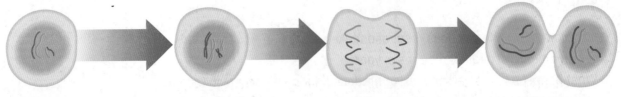

Parent cell with two
pairs of chromosomes

Each chromosome
replicates itself

The copies are pulled apart. Cell
now divides for the only time

Two 'daughter' cells
are formed

Meiosis

Meiosis occurs in the **testes** and **ovaries** to produce the gametes (sperm and egg) for sexual reproduction.

Remember, **gametes** only contain one set of 23 chromosomes.

(HT) When a cell divides to form gametes:
- Copies of the genetic material are made.
- The cell then divides twice to form four gametes, each with a single set of chromosomes.

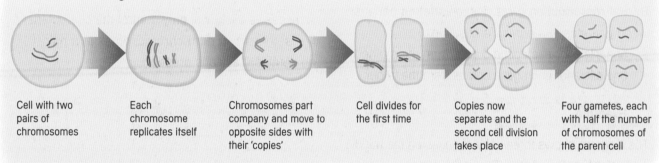

| Cell with two pairs of chromosomes | Each chromosome replicates itself | Chromosomes part company and move to opposite sides with their 'copies' | Cell divides for the first time | Copies now separate and the second cell division takes place | Four gametes, each with half the number of chromosomes of the parent cell |

Inheritance of the Sex Chromosome

Of the 23 pairs of chromosomes in the human body, one pair carries the genes that determine sex:
- **In females**, the **sex chromosomes** are identical and are called the X chromosomes.
- **In males**, one sex chromosome (Y) is much shorter than the other (X).

Female Sex Chromosomes	Male Sex Chromosomes
X X	X Y

As with all chromosomes, offspring inherit...
- one sex chromosome from the mother (always an X chromosome)
- one sex chromosome from the father (either an X or Y chromosome).

So, the sex of the offspring is decided by whether the ovum (egg) is fertilised by an **X-carrying sperm** or a **Y-carrying sperm**.

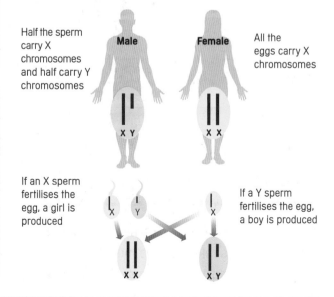

Half the sperm carry X chromosomes and half carry Y chromosomes

All the eggs carry X chromosomes

If an X sperm fertilises the egg, a girl is produced

If a Y sperm fertilises the egg, a boy is produced

Quick Test

1. What type of cell division produces new cells that are identical to the parent cells?
2. What is the combination of sex chromosomes in a male liver cell?
3. What type of cell division produces gametes?
4. How many chromosomes do human gametes contain?

Alleles

Some characteristics are controlled by a single **gene**. **Genes** may have different forms, or variations, called **alleles**. For example…

- the gene that controls tongue-rolling ability has two alleles – either you can or you can't
- the gene that controls eye colour has two alleles – blue or brown.

Sexual reproduction gives rise to **variation** because when gametes join during fertilisation…

- one allele for each gene comes from the mother
- one allele for each gene comes from the father.

In a pair of chromosomes, the alleles for a gene can be the **same** or **different**. If they are different…

- one allele will be **dominant**
- one allele will be **recessive**.

A **dominant** allele **will always control** the characteristic; it will express itself even if present on only one chromosome in a pair.

A recessive allele will **only** control the characteristic if it is present on **both chromosomes in a pair** (i.e. no dominant allele is present).

Example

The diagram shows three pairs of genes on the middle of a pair of chromosomes. The genes code for…

- tongue-rolling ability
- eye colour
- type of earlobe (i.e. attached or unattached).

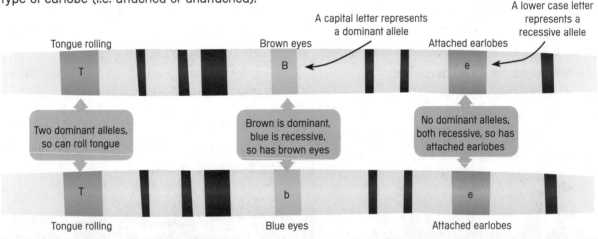

	Homozygous Dominant	Heterozygous	Homozygous Recessive
Tongue rolling	TT (can roll)	Tt (can roll)	tt (can't roll)
Eye colour	BB (brown)	Bb (brown)	bb (blue)
Earlobes	EE (free earlobes)	Ee (free earlobes)	ee (attached earlobes)

Monohybrid Inheritance

When a characteristic is determined by **just one pair** of **alleles** it is referred to as **monohybrid inheritance**. A simple **genetic cross diagram** is a

biological model and can be used to predict the outcome of crosses in monohybrid inheritance.

Genetic Diagrams

In genetic diagrams you should use…
- **capital** letters for **dominant** alleles
- **lower case** letters for **recessive** alleles.

So, for eye colour…
- B is used for brown eye alleles (dominant)
- b is used for blue eye alleles (recessive).

From the crosses on the diagrams, the following can be seen:

1 If one parent has **two dominant** alleles, then **all offspring** inherit the characteristic.

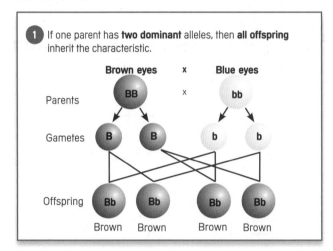

2 If two parents have **one recessive allele each**, then the characteristic **may** appear in offspring (25% chance).

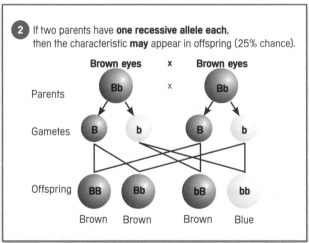

3 If one parent has **one recessive** allele and one parent has **two recessive** alleles, then there is a 50% chance that the characteristic will appear in offspring.

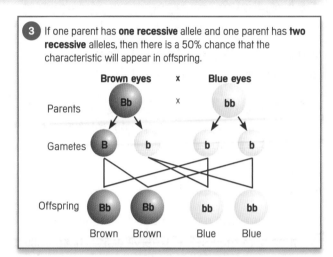

Remember, these are **only probabilities**. In practice, what matters is which egg is fertilised by which sperm. This process is completely **random**.

HT When constructing genetic diagrams remember…
- to clearly identify the alleles of the parents
- to place each of these alleles in a separate gamete
- to join each gamete with the **two gametes** from the other parent.

You should know the following genetic terms:
- **Genotype** – the combination of alleles that an individual has for a particular gene.
- **Homozygous** – an individual who carries two copies of the **same** allele for a particular gene, e.g. **BB** or **bb**.
- **Heterozygous** – an individual who carries two **different** alleles for a particular gene, e.g. **Bb**.
- **Phenotype** – the expression of the genotype (the characteristic shown), e.g. the **homozygous recessive genotype** of bb would have a **phenotype** of blue eyes.

Mendel

Gregor **Mendel** is known as the father of modern genetics. Mendel…
- was an Austrian monk
- studied monohybrid inheritance using pea plants
- proposed the idea of separately inherited factors.

The importance of Mendel's research wasn't recognised until after his death, when other scientists discovered chromosomes and linked them to his ideas.

B2 Cell Division and Inheritance

Stem Cells

Most types of **animal cells** differentiate and become specialised at an early stage. Many **plant cells** retain the ability to differentiate throughout their life.

Stem cells have the ability to develop into any kind of cell because they have not yet **differentiated**. Stem cells in humans are found in…

- human embryos
- adult bone marrow.

Human stem cells can be made to differentiate into any time of human cell, e.g. nerve cells. These cells could potentially be used to help treat people with conditions like paralysis.

There are some **ethical issues** surrounding the use of stem cells for scientific research:

- **Against** – some people believe that human embryos shouldn't be used because they are a potential human life.
- **For** – some people believe it is more important to help people who are already living.

Some embryo stem cells are taken from unwanted embryos from fertility clinics where they would be destroyed if not used for research. Some countries have banned stem cell research. Other countries, including the UK, follow very strict guidelines.

The Structure of Chromosomes

Chromosomes are made up of **DNA** (deoxyribo nucleic acid). A DNA molecule consists of **two long strands** that are **coiled** to form a **double helix**.

Each person has unique DNA (apart from identical twins). So, DNA can be used for identification, i.e. DNA 'fingerprinting'. A **gene** is a small section of DNA.

(HT) Genes code for a particular characteristic by providing a code for a combination of amino acids that make up a specific protein.

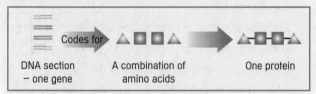

| DNA section – one gene | Codes for | A combination of amino acids | One protein |

A Cell

Chromosomes

A Section of Chromosome

Gene

A Section of Uncoiled DNA

A Section of DNA

Genetic Disorders

Some disorders are inherited, such as…

- **polydactyly**
- **cystic fibrosis**.

People with **polydactyly** have extra fingers or toes. It is caused by a **dominant** allele, so it can be passed on even if only one parent has the disorder.

Cystic fibrosis is a disorder of the cell membranes. It's caused by a **recessive** allele, so both parents must carry the allele. Because the allele is recessive, the parents can be carriers without having the disorder themselves.

Embryos can be screened for the alleles that cause genetic disorders. This has economic, social and ethical issues.

Stem cell • DNA • Polydactyly • Cystic fibrosis

Fossils

Fossils…
- are the **remains of plants or animals** from many years ago
- are found in rocks
- provide evidence of how organisms have changed over time.

Fossils may be formed in various ways:
- From the **hard parts** of animals that **do not decay** easily.
- From parts of organisms that have not **decayed** because one or more of the **conditions** needed for decay are **absent**.
- When parts of the organisms are replaced **by other materials** as they decay.
- As **preserved traces** of organisms, e.g. footprints, burrows and rootlet traces.

Fossils are quite **rare** because…
- many early forms of life were soft-bodied and **did not leave traces behind**
- fossils have been destroyed by geological activity (weathering and erosion).

We can learn from fossils how much or how little different organisms have changed as life has developed on Earth.

Scientists can't be certain how life on Earth began due to a lack of enough valid and reliable evidence.

Extinction of Species

The reason for the **extinction** of a species may include…
- changes to the environment
- new predators
- new diseases
- new, more successful competitors
- a single catastrophic event, e.g. massive volcanic eruptions or collisions with asteroids
- the cyclical nature of speciation.

New Species

New species arise as a result of **isolation**, i.e. two populations of a species become separated, e.g. geographically by a river or a mountain range.

(HT) New species also arise as a result of…
- **genetic variation** – each population has a wide range of alleles that control their characteristics
- **natural selection** – within each population, the alleles that control the characteristics that help the organism to survive are selected
- speciation – the populations become so different that successful interbreeding is no longer possible.

Quick Test

1. What are different forms of genes called?
2. Where are human stem cells found?
3. Does a recessive or dominant allele cause cystic fibrosis?
4. Do females have XX or XY on their sex chromosomes?
5. What is a gene and what is it made from?
6. Why are fossils rare?
7. (HT) What do we call a combination of one dominant and one recessive allele?

B2 Exam Practice Questions

1 All cells contain a nucleus, cytoplasm and cell membrane.

    a) List three other parts that a cell found in a plant leaf would contain.

        chloroplasts , vacuole. **(1 mark)**

    b) What do mitochondria release?

        chemical reactions. **(1 mark)**

2 Photosynthesis is an important reaction in plants.

    a) Complete the word equation below for photosynthesis.

        carbon dioxide + *water* $\xrightarrow[\text{chlorophyll}]{\text{light}}$ **glucose** + *oxygen*

        (2 marks)

    b) Where is chlorophyll found in a plant cell?

        chloroplasts **(1 mark)**

    c) The glucose produced in photosynthesis is often changed into insoluble starch to be stored or used in respiration. Give two other uses for glucose in plants.

        used to produce proteins and used to produce cellulose to strengthen cell walls. **(2 marks)**

3 A scientist wanted to investigate the number of different species found on a hillside.

    a) What piece of equipment is the scientist likely to use to help him sample the area?

        Quadrats **(1 mark)**

    b) Describe how the scientist could carry out his investigation.

        Place the quadrats across the hillside each one the same distance apart and then count how many different species apear in each one.

        (3 marks)

    c) A colleague suggests to the scientist that he should have investigated the effect of the slope on the number of species found on the hillside. Suggest a sampling technique that the scientist might use.

        sampling along a transect. **(1 mark)**

4 Explain what diffusion is.

    The spreading of the particles of a gas or a substance in a solution from a region of high concentration **(1 mark)** *to a region of low concentration.*

5 There are numerous enzymes involved in the digestion of food. Complete the following table about digestive enzymes. **(5 marks)**

Enzyme	Where is it Produced?	What does it Digest?	What does it Produce?
Amylase	Salivary glands, pancreas and small intestine	Starch	*Glucose*
Protease	*Stomach*, pancreas and small intestine	~~All~~ *Proteins*	Amino acids
Lipase	Pancreas and small intestine	Lipids (fats and oils)	Glycerol and *amino acids.*

6 An athlete is training for the Olympics. Describe the changes you would expect to take place in her body as she carries out her training. **(3 marks)**

1. *Heart rate increases*
2. *Bloodflow to muscles increases.*
3. *Arteries supplying muscles dilate*

7 Complete the following sentences by filling in the missing numbers. **(2 marks)**

Human body cells contain __46__ chromosomes. Gametes only contain __23__ chromosomes.

8 Give two reasons why a species may become extinct.

Predictors or change in environment

(2 marks)

HT 9 When an athlete is training, their muscle cells do not always receive enough oxygen to carry out respiration.

a) What reaction happens in the cells that still release energy?

(1 mark)

b) This reaction causes a build up of lactic acid in the muscle cells. Give the term used to describe the amount of oxygen owed to the body and state what happens to the lactic acid.

(2 marks)

C2 The Structure of Substances

Ionic Compounds

Ionic compounds are giant structures of **ions**. They are held together by **strong forces** of attraction (electrostatic forces) between **oppositely charged ions**, that act in **all directions**. This type of bonding is called **ionic bonding**.

Ionic compounds...

*   have **high melting** and **boiling points**
*   **conduct electricity** when molten or in solution because the charged ions are free to move about and carry the current.

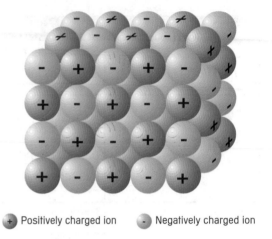

+ Positively charged ion   − Negatively charged ion

The Ionic Bond

An **ionic bond** occurs between a **metal** and a **non-metal**. It involves a **transfer** of **electrons** from one atom to the other.

This forms electrically charged **ions**, each of which has a complete outer energy level.

Ions have the electronic structure of a noble gas.

*   Atoms that **lose electrons** become **positively charged ions**.
*   Atoms that **gain electrons** become **negatively charged ions**.

Example 1
Sodium (Na) and chlorine (Cl) bond ionically to form sodium chloride, NaCl.

1. The sodium atom has one electron in its outer shell.
2. The electron is transferred to the chlorine atom.
3. Both atoms now have eight electrons in their outer shell.
4. The atoms become ions, Na$^+$ and Cl$^-$.
5. The compound formed is sodium chloride, NaCl.

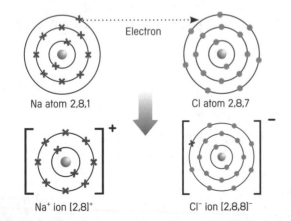

Example 2
Magnesium (Mg) and oxygen (O) bond ionically to form magnesium oxide, MgO.

1. The magnesium atom has two electrons in its outer shell.
2. These two electrons are transferred to the oxygen atom.
3. Both atoms now have eight electrons in their outer shell.
4. The atoms become ions, Mg^{2+} and O^{2-}.
5. The compound formed is magnesium oxide, MgO.

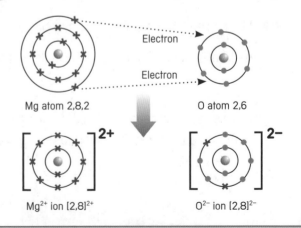

Key Words   **Ionic compound • Ion • Ionic bond • Electron**

Alkali Metals and Halogens

The Alkali Metals (Group 1)

The alkali metals...

- have one electron in their outermost shell
- react with **non-metal elements** to form **ionic compounds** where the metal ion has a single **positive** charge.

Lithium Atom
2,1

Sodium Atom
2,8,1

Potassium Atom
2,8,8,1

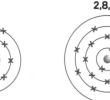

The Halogens (Group 7)

The halogens...

- have seven electrons in their outermost shell
- react with **alkali metals** to form **ionic compounds** where the halide ions have a single **negative** charge.

Fluorine Atom
2,7

Chlorine Atom
2,8,7

Bromine Atom
2,8,8,7

Mixtures and Compounds

A **mixture** consists of two or more elements or compounds that are **not chemically combined**. The properties of the substances remain unchanged and specific to that substance.

Compounds are substances in which the atoms of two or more elements **are chemically combined** (not just mixed together).

Atoms can form chemical bonds by...

- **sharing electrons (covalent bonds)**
- **gaining** or **losing electrons (ionic bonds)**.

When atoms form **chemical bonds**, the arrangement of the **outer shell** of electrons **changes**. This results in each atom getting a **complete outer shell** of electrons. For most atoms this is eight electrons, but for helium it is only two.

Simple Molecular Compounds

Substances that consist of **simple molecules** are gases, liquids and solids that have relatively **low melting** and **boiling points**. The molecules have no overall electrical charge, so they can't conduct electricity.

HT Simple molecular compounds have low melting and boiling points because they have weak intermolecular forces (forces between their molecules).

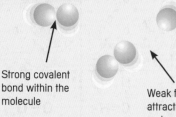

Strong covalent bond within the molecule

Weak forces of attraction between molecules

Quick Test

1. What is a compound?
2. When atoms share a pair of electrons, what type of bond is formed?
3. a) Fill in the missing words:
   Simple covalent substances have relatively low _____ _____ and they have weak _____ forces between the molecules.
   b) When molten or dissolved in water, ionic compounds conduct _____ because the ions are free to _____ .

C2 The Structure of Substances

The Covalent Bond

A **covalent bond** occurs between **non-metal atoms**. It is a strong bond formed when **pairs of electrons are shared**.

Some covalently bonded substances have **simple structures**, e.g. H_2, Cl_2, O_2, HCl, H_2O and CH_4.

Others have **giant covalent structures**, called **macromolecules**, e.g. diamond and silicon dioxide.

Atoms that share electrons usually have **low melting** and **boiling points**. This is because they often form molecules in which there are...

* **strong covalent bonds** between the **atoms**
* **weak forces of attraction** between the **molecules**.

These forces are very weak compared to the strength of covalent bonds.

Example

1. A chlorine atom has seven electrons in its outer shell.
2. In order to bond with another chlorine atom, an electron from each atom is shared.
3. This gives each chlorine atom eight electrons in the outer shell.
4. Each atom now has a complete outer shell.

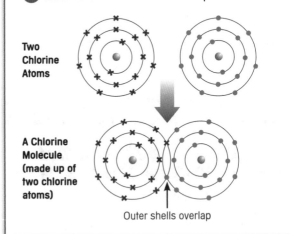

Two Chlorine Atoms

A Chlorine Molecule (made up of two chlorine atoms)

Outer shells overlap

Covalent Bonding

There are three different methods for representing the covalent bonds in each molecule. You need to be familiar with the following examples, and know how to use the different methods.

The two most common forms are shown in the table below.

The third form is shown here for an ammonia molecule. But, unless specifically asked for, you should use the other two methods.

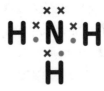

Molecule	Water H_2O	Chlorine Cl_2	Hydrogen H_2	Hydrogen chloride, HCl	Methane CH_4	Oxygen O_2
Method 1	H O H	Cl Cl	H H	H Cl	H C H (with H top and bottom)	O O
Method 2	H–O–H	Cl–Cl	H–H	H–Cl	H–C–H (with H top and bottom)	O=O (a double bond)

Giant Covalent Structures

All the atoms in giant covalent structures are linked by **strong covalent bonds**. This means they have very **high melting points**.

Diamond is a form of carbon that has a **giant, rigid covalent structure** (lattice). Each carbon atom forms **four covalent bonds** with other carbon atoms. Diamond has a **large number of covalent bonds** so it's a very **hard substance**.

Graphite is a form of carbon that also has a giant covalent structure. However, in graphite, each carbon atom forms **three covalent bonds** with other carbon atoms in a layered structure. Graphite has layers that can slide past each other, making it soft and slippery.

HT The layers in graphite are held together by weak intermolecular forces. In graphite, one electron from each carbon atom is **delocalised**. These delocalised electrons allow graphite to **conduct heat** and **electricity**.

Silicon dioxide (or silica, SiO_2) has a lattice structure similar to diamond. Each **oxygen** atom is joined to **two silicon atoms**, and each **silicon** atom is joined to **four oxygen atoms**.

Diamond	Graphite
Covalent bond between carbon atoms ● Carbon atom	Covalent bond between carbon atoms ● Carbon atom · Weak bond between layers

Silicon Dioxide
Covalent bond ◐ Silicon atom ● Oxygen atom

HT Fullerenes

Carbon can also form molecules known as **fullerenes**, which have different numbers of carbon atoms.

The structure of fullerenes is based on hexagon rings of carbon atoms.

Fullerenes can be used to deliver drugs in the body, in lubricants, as catalysts, and in **nanotubes** for reinforcing materials, e.g. in tennis rackets.

Metals

The layers of atoms in metals are able to slide over each other. This means that metals can be **bent and shaped**.

HT Metals have a giant structure in which electrons in the highest energy level can be delocalised. This produces a regular arrangement (lattice) of **positive ions** that are held together by electrons using electrostatic attraction.

These delocalised electrons can move around freely. This allows metals to **conduct heat and electricity**.

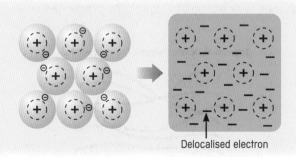

Delocalised electron

C2 The Structure of Substances

Alloys

An **alloy** is a mixture that contains a **metal** and at least one other **element**.

The added element disturbs the regular arrangement of the metal atoms so the layers don't slide over each other so easily. So, **alloys** are usually stronger and harder than pure metal.

Shape Memory Alloys

Go back to original shape

Smart alloys belong to a group of materials that are being developed to meet the demands of modern engineering and manufacturing. These materials respond to changes in their environment.

Smart alloys **remember their shape.** They can be deformed but will return to their original shape, for example, flexible spectacle frames. Nitinol (used in dental braces) is an example of a smart alloy.

Properties of Polymers

The properties of polymers depend on…
*   what they're made from (i.e. what **monomer** is used)
*   the conditions (i.e. temperature and catalyst) under which they're made.

For example, low density poly(ethene) (LDPE) and high density poly(ethene) (HDPE) are both made from the monomer ethene. But the polymers have different properties because different catalysts and reaction conditions are used to make them.

LDPE is used to make carrier bags and HDPE is used to make plastic bottles.

Thermo-softening and Thermo-setting Polymers

Thermo-softening polymers consist of individual polymer chains that are tangled together (like spaghetti).

(HT) There are weak intermolecular forces between all of the polymer chains in a thermo-softening polymer, which helps to explain its properties.

Thermo-setting polymers consist of polymer chains that are joined together by cross-links between them. Thermo-setting polymers don't melt when they're heated.

Thermo-softening Polymer (no cross-links)

Thermo-setting Polymer

Cross-links

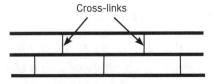

Key Words   Smart alloy • Monomer • Thermo-softening • Thermo-setting

Nanoparticles and Nanostructures

Nanoscience is the study of very small structures. The structures are 1–100 nanometres in size, roughly in the order of a few hundred atoms.

One **nanometre** is 0.000 000 001m (one billionth of a metre) and is written as 1nm or 1×10^{-9}m.

Nanoparticles are tiny, tiny particles that can combine to form structures called **nanostructures**.

Nanostructures can be manipulated so materials can be developed that have new and specific properties.

The **properties** of **nanoparticles** are different to the properties of the **same materials in bulk**.

For example...
- nanoparticles are more sensitive to light, heat and magnetism
- nanoparticles have a high surface area in relation to their volume.

Research into nanoparticles may lead to the development of new...
- computers
- catalysts
- coatings
- highly selective sensors
- stronger and lighter construction materials
- cosmetics, e.g. suntan creams and deodorants.

Human Hair	Virus / Small Bacteria	Atoms and Small Molecules
0.000 01metre = 1×10^{-5}m	0.000 000 1metre = 1×10^{-7}m	0.000 000 001metre = 1×10^{-9}m
Can be seen using a microscope	Can be seen using an electron microscope	Nanoparticle zone

Quick Test

1. Why are metals able to be bent and shaped?
2. What is the main structural difference between thermo-softening and thermo-setting polymers?
3. Give two potential uses of nanoparticles.

The Periodic Table

1	2											3	4	5	6	7	8 or 0
						1 H hydrogen 1											4 He helium 2
7 Li lithium 3	9 Be beryllium 4											11 B boron 5	12 C carbon 6	14 N nitrogen 7	16 O oxygen 8	19 F fluorine 9	20 Ne neon 10
23 Na sodium 11	24 Mg magnesium 12											27 Al aluminium 13	28 Si silicon 14	31 P phosphorus 15	32 S sulfur 16	35.5 Cl chlorine 17	40 Ar argon 18
39 K potassium 19	40 Ca calcium 20	45 Sc scandium 21	48 Ti titanium 22	51 V vanadium 23	52 Cr chromium 24	55 Mn manganese 25	56 Fe iron 26	59 Co cobalt 27	59 Ni nickel 28	63.5 Cu copper 29	65 Zn zinc 30	70 Ga gallium 31	73 Ge germanium 32	75 As arsenic 33	79 Se selenium 34	80 Br bromine 35	84 Kr krypton 36
85 Rb rubidium 37	88 Sr strontium 38	89 Y yttrium 39	91 Zr zirconium 40	93 Nb niobium 41	96 Mo molybdenum 42	[98] Tc technetium 43	101 Ru ruthenium 44	103 Rh rhodium 45	106 Pd palladium 46	108 Ag silver 47	112 Cd cadmium 48	115 In indium 49	119 Sn tin 50	122 Sb antimony 51	128 Te tellurium 52	127 I iodine 53	131 Xe xenon 54
133 Cs caesium 55	137 Ba barium 56	139 La* lanthanum 57	178 Hf hafnium 72	181 Ta tantalum 73	184 W tungsten 74	186 Re rhenium 75	190 Os osmium 76	192 Ir iridium 77	195 Pt platinum 78	197 Au gold 79	201 Hg mercury 80	204 Tl thallium 81	207 Pb lead 82	209 Bi bismuth 83	[209] Po polonium 84	[210] At astatine 85	[222] Rn radon 86
[223] Fr francium 87	[226] Ra radium 88	[227] Ac* actinium 89	[261] Rf rutherfordium 104	[262] Db dubnium 105	[266] Sg seaborgium 106	[264] Bh bohrium 107	[277] Hs hassium 108	[268] Mt meitnerium 109	[271] Ds darmstadtium 110	[272] Rg roentgenium 111							

N.B. The exact position of the mass number/relative atomic mass, element name and atomic number may differ depending on the version of Periodic Table. However, the mass number will always be the larger number, and the atomic number the smaller number.

Mass Number and Atomic Number

Atoms of an element can be described using their **mass number** and **atomic number**.

The **mass number** is the total number of **protons** and **neutrons** in the atom.

The **atomic (proton) number** is the number of protons in the atom.

Mass number → 23 **Na** ← Element symbol

Atomic number → 11 **Na**

Top

Bottom.

Number of neutrons = Mass number — Atomic number

The number of **protons** in an atom is **equal** to the number of **electrons**. So, an atom has **no overall charge**.

Examples

Hydrogen

$_1^1\text{H}$

1 proton
1 electron

Oxygen

$_8^{16}\text{O}$

8 protons
8 electrons

Although they have the same charge, protons and electrons have a **different mass**.

Atomic Particle	Relative Mass
Proton 🔴	1
Neutron ⚪	1
Electron ✖	Very small (negligible)

Isotopes

All atoms of a **particular element** have the **same number** of protons. Atoms of **different elements** have **different numbers** of protons.

Isotopes are atoms of the **same element** that have **different numbers of neutrons.**

Isotopes have the **same atomic number** but a **different mass number**.

For example, chlorine has two isotopes.

$$^{35}_{17}Cl \qquad ^{37}_{17}Cl$$

17 protons
17 electrons
18 neutrons (35 − 17)

17 protons
17 electrons
20 neutrons (37 − 17)

Relative Atomic Mass, A_r

The **relative atomic mass**, A_r, of an element is found on the Periodic Table. It is the larger number shown for each element.

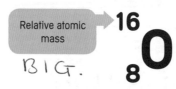

Relative atomic mass

$$^{16}_{8}O$$

BIG.

(HT) The relative atomic mass, A_r, is the mass of a particular atom compared with a twelfth of the mass of a carbon atom (the ^{12}C isotope).

The A_r is an **average** value for all the **isotopes** of the element.

By looking at the Periodic Table, you can see that...
- carbon is 12 times heavier than hydrogen, but is only half as heavy as magnesium
- magnesium is three-quarters as heavy as sulfur
- sulfur is twice as heavy as oxygen, etc.

You can use this idea to calculate the **relative formula mass** of compounds.

Relative Formula Mass, M_r

The **relative formula mass**, M_r, of a compound is the relative atomic masses of all its elements added together.

To calculate M_r, you need to know...
- the formula of the compound
- the A_r of all the atoms involved.

Example 1
Calculate the M_r of water, H_2O.

Write the formula

Substitute the A_rs

Calculate the M_r

$$H_2O$$

$(2 \times 1) + 16$

$2 + 16 = 18$

Example 2
Calculate the M_r of potassium carbonate, K_2CO_3.

Write the formula

Substitute the A_rs

Calculate the M_r

$$K_2CO_3$$

$(39 \times 2) + 12 + (16 \times 3)$

$78 + 12 + 48 = 138$

Quick Test

1. What is the total number of protons and neutrons in an atom called?
2. What are atoms of the same element that have different numbers of neutrons called?

Calculating Percentage Mass

The <u>mass of the compound is its</u> **relative formula mass** in grams.

To calculate the **percentage mass** of an element in a compound, you need to know...
- the **formula** of the compound
- the **relative atomic mass** of all the atoms.

You can calculate the percentage mass by using this formula:

$$\frac{\text{Relative mass of element in the compound}}{\text{Relative formula mass of compound } (M_r)} \times 100$$

Example 1
Calculate the percentage mass of magnesium in magnesium oxide, MgO.

magnesium   oxygen

Relative mass of magnesium = 24
Relative formula mass (M_r) of MgO =
$$24 + 16 = 40$$

A_r Mg A_r O M_r MgO

$$\frac{\text{Relative mass of element}}{M_r \text{ of compound}} \times 100$$

$$= \frac{24}{40} \times 100 = \textbf{60\%}$$

Example 2
Calculate the percentage mass of potassium in potassium carbonate, K_2CO_3.

potassium   carbon   oxygen

Relative mass of potassium = 39 x 2
Relative formula mass (M_r) of K_2CO_3 =
$$78 + 12 + 48 = 138$$

A_r K x 2 A_r C A_r O x 3 M_r K_2CO_3

$$\frac{\text{Relative mass of element}}{M_r \text{ of compound}} \times 100$$

$$= \frac{78}{138} \times 100 = \textbf{56.5\%}$$

HT Empirical Formula of a Compound

The empirical formula of a compound is the **simplest whole number ratio** of each kind of atom in the compound.

Example
Find the simplest formula of an oxide of iron produced by reacting 1.12g of iron with 0.48g of oxygen (A_r Fe = 56, A_r O = 16).

Identify the mass of the elements in the compound

Masses: Fe = 1.12, O = 0.48

Divide these masses by their relative atomic masses

$$Fe = \frac{1.12}{56} = 0.02 \qquad O = \frac{0.48}{16} = 0.03$$

Identify the ratio of atoms in the compound

Ratio =   0.02 : 0.03
x 100 ⟶ 2 : 3 ⟵ x 100

Empirical formula = **Fe_2O_3**

The Mole

A **mole** (mol) is a measure of the **number of particles** (atoms or molecules) contained in a substance. One mole of a substance is its relative formula mass or A_r in grams.

One mole of **any substance** (element or compound) will always contain the **same number** of particles – six hundred thousand billion billion or 6×10^{23}. This is the **relative formula mass** of the substance.

If a substance is an **element**, the mass of one mole of the substance, called the molar mass (g/mol), is always **equal** to the **relative atomic mass** of the substance in grams. For example...

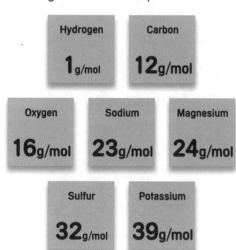

Hydrogen	Carbon
1g/mol	**12**g/mol

Oxygen	Sodium	Magnesium
16g/mol	**23**g/mol	**24**g/mol

Sulfur	Potassium
32g/mol	**39**g/mol

If a substance is a **compound**, the mass of one mole of the substance is always **equal** to the **relative formula mass** of the substance in grams. For example, one mole of sodium hydroxide (NaOH)...

A_r sodium + A_r hydrogen + A_r oxygen

$$= 23 + 1 + 16 = \textbf{40g}$$

You can calculate the number of moles in a substance using this formula:

$$\text{Number of moles of substance (mol)} = \frac{\text{Mass of substance (g)}}{\text{Mass of one mole (g/mol)}}$$

N.B. You need to remember this equation as it will not be given to you in the exam.

Example 1

Calculate the number of moles of carbon in 36g of the element.

$$\text{Number of moles of substance (mol)} = \frac{\text{Mass of substance (g)}}{\text{Mass of one mole (g/mol)}}$$

$$= \frac{36g}{12g/mol} \quad \text{←} \quad \boxed{A_r \text{ carbon} = 12}$$

$$= \textbf{3 moles}$$

Example 2

Calculate the number of moles of carbon dioxide in 33g of the gas.

$$\text{Number of moles of substance (mol)} = \frac{\text{Mass of substance (g)}}{\text{Mass of one mole (g/mol)}}$$

$$= \frac{33g}{44g/mol} \quad \text{←} \quad \boxed{\begin{array}{l} A_r \text{ carbon dioxide} \\ = A_r \text{ carbon +} \\ 2 \times A_r \text{ oxygen} \\ = 12 + (2 \times 16) \\ = 44 \end{array}}$$

$$= \textbf{0.75 mole}$$

Example 3

Calculate the mass of four moles of sodium hydroxide.

$$\text{Mass of substance (g)} = \text{Number of moles of substance (mol)} \times \text{Mass of one mole (g/mol)}$$

$$= \text{4mol} \times \text{40g/mol}$$

$$= \textbf{160g}$$

N.B. If you are confident in your mathematical ability, you can also do these calculations using ratios.

Quick Test

1. Calculate the percentage by mass of nitrogen (N) in each of the following compounds:
   a) HNO_3 **b)** NH_4NO_3 **c)** $(NH_4)_2SO_4$
2. Calculate the number of moles of the following:
   a) 48g of magnesium **b)** 12.4g of SO_2 gas
3. Calculate the mass of the following:
   a) Two moles of carbon monoxide (CO)
   b) Three moles of potassium carbonate (K_2CO_3)

HT Calculating the Mass of a Product

Example

Calculate how much calcium oxide can be produced from 50kg of calcium carbonate. (Relative atomic masses: Ca = 40, C = 12, O = 16).

① Write down the equation.

② Work out the M_r of each substance.

③ Check that the total mass of reactants equals the total mass of the products. If they are not the same, check your work.

④ The question only mentions calcium oxide and calcium carbonate, so you can now ignore the carbon dioxide. You just need the ratio of mass of reactant to mass of product.

⑤ Use the ratio to calculate how much calcium oxide can be produced.

①
$$CaCO_3 \rightarrow CaO + CO_2$$

②
$$40 + 12 + (3 \times 16) \rightarrow (40 + 16) + [12 + (2 \times 16)]$$

③
$$100 \rightarrow 56 + 44 \ ✔$$

④
$$100 : 56$$

⑤
If 100kg of $CaCO_3$ produces 56kg of CaO, then 1kg of $CaCO_3$ produces $\frac{56}{100}$ kg of CaO, and 50kg of $CaCO_3$ produces $\frac{56}{100} \times 50$

= 28kg of CaO

HT Calculating the Mass of a Reactant

Example

Calculate how much aluminium oxide is needed to produce 540 tonnes of aluminium. (Relative atomic masses: Al = 27, O = 16).

① Write down the equation.

② Work out the M_r of each substance.

③ Check that the total mass of reactants equals the total mass of the products. If they are not the same, check your work.

④ The question only mentions aluminium oxide and aluminium, so you can now ignore the oxygen. You just need the ratio of mass of reactant to mass of product.

⑤ Use the ratio to calculate how much aluminium oxide is needed.

①
$$2Al_2O_3 \rightarrow 4Al + 3O_2$$

②
$$2 \times [(2 \times 27) + (3 \times 16)] \rightarrow (4 \times 27) + [3 \times (2 \times 16)]$$

③
$$204 \rightarrow 108 + 96 \ ✔$$

④
$$204 : 108$$

⑤
If 204 tonnes of Al_2O_3 produces 108 tonnes of Al, then $\frac{204}{108}$ tonnes is needed to produce 1 tonne of Al, and $\frac{204}{108} \times 540$ tonnes is needed to produce 540 tonnes of Al

= 1020 tonnes of Al_2O_3

Instrumental Methods

Standard laboratory equipment can be used to detect and **identify elements** and **compounds**. But **instrumental methods** that involve using highly **accurate** instruments to analyse and identify substances have been developed to perform this function.

These instruments give rapid results, are very sensitive and accurate, and can be used on small samples.

An example of a common instrumental method of analysis is Gas Chromatography linked to Mass Spectroscopy (GC-MS).

A GC-MS works by allowing different substances, carried by a gas, to travel through a column packed with solid material at different speeds so that they separate out. Each substance will produce a separate peak on an output known as a gas chromatograph. The number of peaks on this output shows the number of compounds present in the original sample.

The position of the peaks on the output graph indicates the retention time, i.e. the time taken to pass through the gas chromatograph.

If the output of the gas chromatography column is linked to a mass spectrometer then this can also be used to identify the substances leaving the column.

(HT) The mass spectrometer can give the relative molecular mass (M_r) of each substance separated in the column.

The molecular mass is given by the molecular ion peak on the spectrum.

Chromatography

Chemical analysis can be used to identify additives in food. **Chromatography** is used to identify artificial colours, by comparing them to known substances.

1. Samples of five known food colourings (A, B, C, D and E), and the unknown substance (X) are put on a 'start line' on a piece of paper.

2. The paper is dipped into a solvent. The solvent dissolves the samples and carries them up the paper.

3. Substance X can be identified by comparing the horizontal spots.

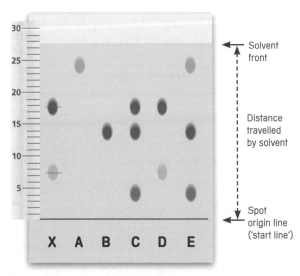

We can see that substance X is food colouring D

Yield

Atoms are **never lost or gained** in a chemical reaction. But, it isn't always possible to obtain the calculated amount of the product for several reasons:

- If the reaction is reversible, it might not go to completion.
- Some product could be lost when it's separated from the reaction mixture.
- Some of the reactants may react in different ways to the expected reaction.

The amount of product obtained is called the **yield**.

The **percentage yield** can be calculated by comparing…

- the actual yield obtained from a reaction
- the maximum theoretical yield.

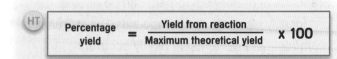

(HT)

$$\text{Percentage yield} = \frac{\text{Yield from reaction}}{\text{Maximum theoretical yield}} \times 100$$

Example

50kg of calcium carbonate ($CaCO_3$) is expected to produce 28kg of calcium oxide (CaO).

A company heats 50kg of calcium carbonate in a kiln and obtains 22kg of calcium oxide.

Calculate the percentage yield.

$$\text{Percentage yield} = \frac{22}{28} \times 100$$

$$= \textbf{78.6\%}$$

Reversible Reactions

Some chemical reactions are **reversible**. In a **reversible reaction**, the **products** can **react** to produce the **original reactants**.

These reactions are represented as…

$$\boxed{A + B \rightleftharpoons C + D}$$

This means that A and B can react to produce C and D, and C and D can also react to produce A and B.

For example…

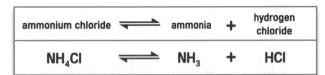

| ammonium chloride $\rightleftharpoons$ | ammonia | + | hydrogen chloride |
| NH_4Cl $\rightleftharpoons$ | NH_3 | + | HCl |

Solid ammonium chloride decomposes when heated to produce ammonia and hydrogen chloride gas (both colourless).

Ammonia reacts with hydrogen chloride gas to produce clouds of white ammonium chloride powder.

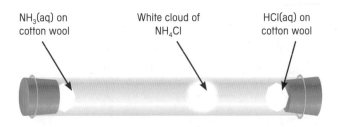

NH_3(aq) on cotton wool   White cloud of NH_4Cl HCl(aq) on cotton wool

Quick Test

1. Give two advantages of using instrumental methods to detect and identify elements.
2. Give two reasons why it may not always be possible to calculate the amount of a product formed in a reaction.
3. (HT) What information does the mass spectrometer give about a compound?
4. (HT) Calculate the mass of carbon dioxide formed if 3g of carbon reacts with oxygen. The equation for the reaction is $C + O_2 \longrightarrow CO_2$ (Relative atomic masses: C = 12, O = 16)

Rates of Reactions

Chemical reactions only occur when reacting particles **collide** with each other with **sufficient energy**.

The **minimum amount** of energy required to cause a reaction is called the **activation energy**.

There are four important factors that affect the rate of reaction:

- Temperature.
- Concentration.
- Surface area.
- Use of a catalyst.

Temperature

In a **cold** reaction mixture the particles move quite **slowly**. They collide less often, with less energy, so **fewer collisions** are successful.

In a **hot** reaction mixture the particles move more **quickly**. They collide more often, with greater energy, so **more collisions** are successful.

Cold Reaction	Hot Reaction

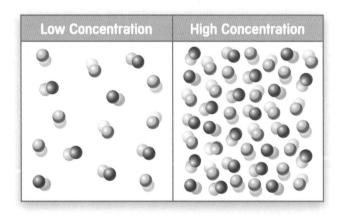

Concentration

In a **low concentration** reaction, the particles are **spread out**. They collide less often, so there are fewer successful collisions.

In a **high concentration** reaction, the particles are crowded **close together**. They collide more often, so there are more successful collisions.

Increasing the **pressure** of reacting gases also increases the frequency of collisions.

Low Concentration	High Concentration

(HT) **Concentrations** of solutions are given in **moles per cubic decimetre (mol/dm³)**.

Equal volumes of solutions of the same molar concentration contain the same number of moles of solute, i.e. the same number of particles.

Equal volumes of gases at the same temperature and pressure contain the **same number** of particles.

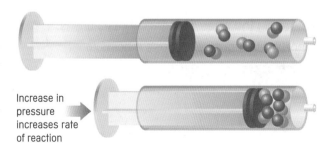

Increase in pressure increases rate of reaction

C2 Rates of Reaction

Surface Area

Large pieces of a solid reactant have a **small surface area** in relation to their volume.

Fewer particles are exposed and available for collisions. This means **fewer collisions** and a **slower reaction**.

Small pieces of a solid reactant have a **large surface area** in relation to their volume, so more particles are exposed and available for collisions.

This means **more collisions** and a **faster reaction**.

Large Pieces

Small Pieces

Using a Catalyst

Speeds up rate of reaction

A **catalyst** is a substance that **changes the rate** of a chemical reaction without being used up or altered in the process.

A catalyst...
- reduces the amount of energy needed for a successful collision
- makes more collisions successful
- speeds up the reaction
- provides a surface for the molecules to attach to, which increases their chances of bumping into each other.

Different reactions need different catalysts. For example...
- the cracking of hydrocarbons uses broken pottery
- the manufacture of ammonia uses iron.

Increasing the rates of chemical reactions is important in industry because it helps to **reduce costs**.

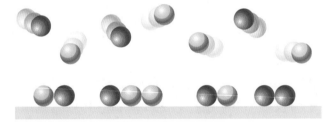

A Catalyst Provides a Larger Surface Area

Catalyst

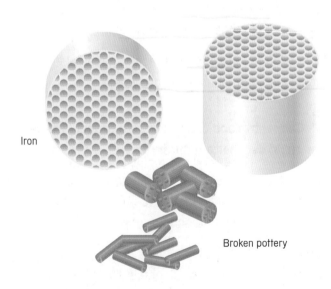

Catalysts Used in Industrial Reactions

Iron

Broken pottery

Analysing the Rate of Reaction

$$\text{Rate of Reaction} = \frac{\text{Amount of reactant used OR product formed}}{\text{Time}}$$

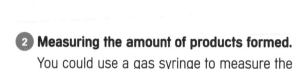

The rate of a chemical reaction can be found in two ways:

1 Measuring the amount of reactants used.

If one of the products is a gas, you could weigh the reaction mixture before and after the reaction takes place. The mass of the mixture will decrease.

2 Measuring the amount of products formed.

You could use a gas syringe to measure the total volume of gas produced at timed intervals.

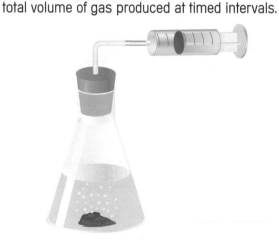

Plotting Reaction Rates

Graphs can be plotted to show the progress of a chemical reaction. There are three things to remember:

1 The steeper the line, the faster the reaction.

2 When one of the reactants is used up, the reaction stops (the line becomes horizontal).

3 The same amount of product is formed from the same amount of reactants, irrespective of rate.

The graph shows us that reaction **A** is faster than reaction **B**. This could be due to several factors, including:

- The surface area of the solid reactants in **A** is greater than in **B**.
- The temperature of reaction **A** is greater than reaction **B**.
- The concentration of the solution in **A** is greater than in **B**.
- A catalyst is used in reaction **A** but not in reaction **B**.

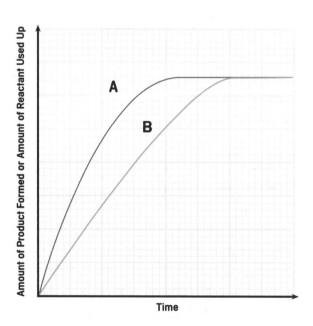

C2 Exothermic and Endothermic Reactions

Chemical Reactions

When chemical reactions occur, **energy** is transferred **to** or **from** the **surroundings**.

Many chemical reactions are, therefore, accompanied by a **temperature change**.

Exothermic Reactions

Exothermic reactions are accompanied by a **temperature rise**. They transfer heat energy to the surroundings, i.e. they **give out** heat.

Common examples of exothermic reactions include…
- neutralising alkalis with acids
- oxidation
- combustion
- self-heating can (e.g. for coffee)
- handwarmers.

methane (natural gas)	+	oxygen	→	carbon dioxide	+	water	+	heat energy

$$CH_4 + 2O_2 \rightarrow CO_2 + 2H_2O$$

carbon	+	oxygen	→	carbon dioxide	+	heat energy

$$C + O_2 \rightarrow CO_2$$

Endothermic Reactions

Endothermic reactions are accompanied by a **fall in temperature**. Heat energy is transferred from the surroundings, i.e. they **take in** heat.

Thermal decomposition and dissolving ammonium nitrate crystals in water are examples of endothermic reactions.

ammonium nitrate	+	water	+	heat energy	→	ammonium nitrate solution

$$NH_4NO_3 + H_2O \rightarrow NH_4NO_3$$

Some sports injury packs are based on endothermic reactions.

Reversible Reactions

If a reversible reaction is exothermic in one direction then it is endothermic in the opposite direction. The same amount of energy is transferred in each case.

Example

hydrated copper sulfate (blue) $\underset{\text{exothermic}}{\overset{\text{endothermic}}{\rightleftharpoons}}$ anhydrous copper sulfate (white) + water

Quick Test

1. How does increasing the temperature increase the rate of a chemical reaction?
2. Fill in the missing words: Catalysts are used in industrial reactions because they _____ the rate of chemical reactions and _____ costs.
3. In an endothermic reaction, where is the energy absorbed from?
4. In a reversible reaction if the forward reaction is exothermic what can you conclude about the backward reaction?

State Symbols

State symbols are used in equations. The symbols are (s) **solid**, (l) **liquid**, (g) **gas** and (aq) **aqueous.**

An **aqueous solution** is produced when a substance is **dissolved in water**.

Soluble Salts from Metals

Metals react with dilute acid to form a **metal salt** and **hydrogen**.

Salt is a word used to describe any metal compound made from a reaction between a metal and an acid.

Some metals react with acid more vigorously than others:

- Silver – no reaction.
- Zinc – fairly reasonable reaction.
- Magnesium – vigorous reaction.
- Potassium – very violent and dangerous reaction.

Soluble Salts from Insoluble Bases

Bases are the oxides and hydroxides of metals. **Soluble** bases are called **alkalis**.

The oxides and hydroxides of transition metals are **insoluble**. Their salts are prepared in the following way:

1. The metal oxide or hydroxide is added to an acid until no more will react.
2. The excess metal oxide or hydroxide is then filtered, leaving a solution of the salt.
3. The salt solution is then evaporated.

This reaction can be written generally as follows:

Example

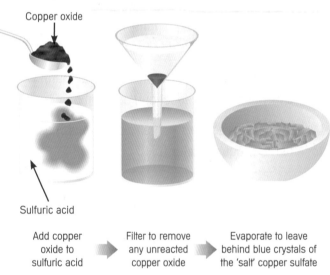

| Add copper oxide to sulfuric acid | Filter to remove any unreacted copper oxide | Evaporate to leave behind blue crystals of the 'salt' copper sulfate |

Salts of Alkali Metals

Compounds of alkali metals, called **salts**, can be made by reacting solutions of their hydroxides (which are alkaline) with a particular acid. This neutralisation reaction can be represented as follows:

The salt produced depends on the metal in the alkali and the acid used.

	Hydrochloric Acid	Sulfuric Acid	Nitric Acid
Metal Hydroxide	Metal chloride	Metal sulfate	Metal nitrate

C2 Acids, Bases and Salts

Insoluble Salts

Insoluble salts can be made by mixing appropriate solutions of ions so that a **precipitate** (solid substance) is formed.

Precipitation can be used to remove unwanted ions from a solution, e.g. softening hard water. The calcium (or magnesium) ions are precipitated out as insoluble calcium (or magnesium) carbonate.

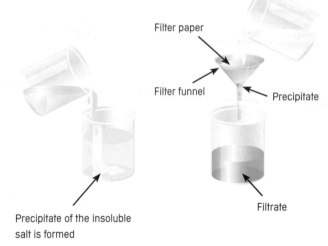

1. Two solutions of soluble substance are mixed together in a beaker

2. The precipitate is filtered off, rinsed and dried

Filter paper

Filter funnel

Precipitate

Filtrate

Precipitate of the insoluble salt is formed

Neutralisation

Acids and alkalis are **chemical opposites**:

- **Acids** contain **hydrogen ions**, $H^+_{(aq)}$.
- **Alkalis** contain **hydroxide ions**, $OH^-_{(aq)}$.

If they are added together in the correct amounts they can 'neutralise' (cancel out) each other.

When an acid reacts with an alkali, the **ions** react together to produce **water** (pH 7).

$$H^+_{(aq)} \quad + \quad OH^-_{(aq)} \quad \longrightarrow \quad H_2O_{(l)}$$

This type of reaction is called **neutralisation** because the solution that remains has a pH of 7, showing it is neutral. For example, hydrochloric acid and potassium hydroxide can be neutralised.

hydrochloric acid	+	potassium hydroxide	→	potassium chloride	+	water
$HCl_{(aq)}$	+	$KOH_{(aq)}$	→	$KCl_{(s)}$	+	$H_2O_{(l)}$

Neutralising HCl and KOH

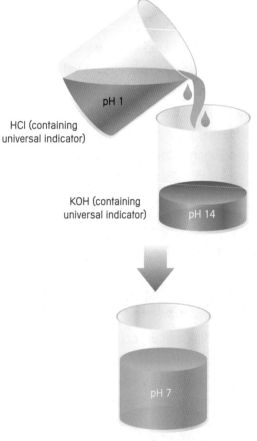

HCl (containing universal indicator)

pH 1

KOH (containing universal indicator)

pH 14

pH 7

KCl + H₂O (containing universal indicator)

Precipitate • Precipitation • Acid • Alkali • Neutralisation

Neutralising Ammonia

Ammonia is an **alkaline gas** that dissolves in water to make an **alkaline solution**.

It's mainly used in the production of fertilisers to increase the nitrogen content of the soil.

Ammonia neutralises nitric acid to produce **ammonium nitrate**. The aqueous ammonium nitrate is then evaporated to dryness.

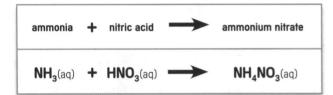

ammonia + nitric acid ⟶ ammonium nitrate

$$NH_3(aq) + HNO_3(aq) \longrightarrow NH_4NO_3(aq)$$

Ammonium nitrate, a fertiliser rich in nitrogen, is also known as 'nitram' (nitrate of ammonia). **Nitrogen-based fertilisers** are important chemicals because they increase the yield of crops.

But nitrates can create problems if they find their way into streams, rivers or groundwater. Nitrates can…
- upset the natural balance of water
- contaminate our drinking water.

Ammonium hydroxide can be neutralised with acids to produce ammonium salts.

	Hydrochloric Acid	Sulfuric Acid	Nitric Acid
Ammonium Hydroxide	Ammonium chloride and water	Ammonium sulfate and water	Ammonium nitrate and water

Indicators and pH Scale

Indicators are dyes that change colour depending on whether they are in **acidic** or **alkaline** solutions:
- **Litmus** is an indicator that changes colour from red to blue or vice versa.
- **Universal indicator** is a mixture of dyes that show a **range** of colours to indicate **how** acidic or alkaline a substance is.

The **pH scale** is a measure of the **acidity or alkalinity** of an **aqueous solution**, across a **14-point scale**. When substances dissolve in water, they dissociate into their individual ions:
- Hydroxide ions, $OH^-(aq)$, make solutions alkaline.
- Hydrogen ions, $H^+(aq)$, make solutions acidic.

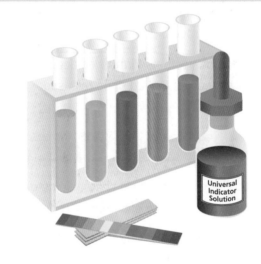

Very acidic   Neutral   Very alkaline

1 2 3 4 5 6 7 8 9 10 11 12 13 14

Quick Test

1. By what kind of reaction are insoluble salts formed?
2. Fill in the missing words:
   a) Soluble hydroxides are called _____ .
   b) Ammonium salts are used as _____ .
   c) In neutralisation reactions H^+ ions react with OH^- to form _____ .

C2 Electrolysis

Electrolysis

LEARN

Electrolysis is the **breaking down** of a compound containing **ions** into its **elements** using an **electrical current**. The substance being broken down is called the **electrolyte.**

Ionic substances are chemical compounds that allow an **electric current** to flow through them when they are…
- molten
- dissolved in water.

These compounds contain **negative** and **positive** ions. During electrolysis…
- **negatively charged ions** move to the **positive electrode**
- **positively charged ions** move to the **negative electrode**.

When this happens, simpler substances are released at the two electrodes.

This moving of electrons forms electrically **neutral** atoms or molecules that are then released.

If there is a **mixture of ions** in the solution, the products formed depend on the reactivity of the elements involved.

For example, in the electrolysis of **copper chloride solution**, the simple substances released are…
- copper at the negative electrode
- chlorine gas at the positive electrode.

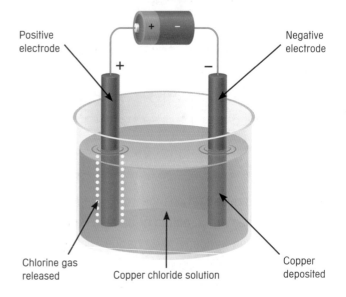

Positive electrode

Negative electrode

Chlorine gas released

Copper chloride solution

Copper deposited

Redox Reactions

Reduction is when **positively** charged ions **gain** electrons at the **negative** electrode.

Oxidation is when **negatively** charged ions **lose** electrons at the **positive** electrode.

A **redox reaction** is a chemical reaction where **both reduction and oxidation** occur.

You can remember this by thinking of the word **oilrig**:
- **O**xidation **I**s **L**oss of electrons (**OIL**).
- **R**eduction **I**s **G**ain of electrons (**RIG**).

LEARN

Electroplating

Electrolysis can be used to electroplate objects with metals such as copper or silver.

Extraction of Aluminium

Aluminium is manufactured by the electrolysis of a molten mixture of aluminium oxide and **cryolite**. The cryolite lowers the melting point of the aluminium oxide, meaning cheaper energy costs.

Aluminium forms at the negative electrode and oxygen gas forms at the positive carbon electrode. The oxygen reacts with the carbon, forming carbon dioxide.

Key Words   Electrolysis • Current • Electrolyte • Electrode • Reduction • Oxidation • Cryolite

Electrolysis of Sodium Chloride Solution

Sodium chloride (common salt) is a compound of an alkali metal and a halogen. It is found in large quantities in the sea and in underground deposits.

Electrolysis of sodium chloride solution (brine) produces some important reagents for the chemical industry:

- **Chlorine gas** (at the positive **electrode**).
- **Hydrogen gas** (at the negative electrode).
- **Sodium hydroxide solution** (passed out of the cell).

Chlorine is used to kill bacteria in drinking water and swimming pools, and to manufacture hydrochloric acid, disinfectants, bleach and PVC.

Hydrogen is used in the manufacture of ammonia and margarine.

Sodium hydroxide is used in the manufacture of soap, paper and ceramics.

Chlorine bleaches damp litmus paper. This is how its presence can be detected in a laboratory.

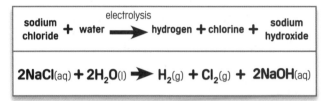

$$2NaCl_{(aq)} + 2H_2O_{(l)} \longrightarrow H_{2(g)} + Cl_{2(g)} + 2NaOH_{(aq)}$$

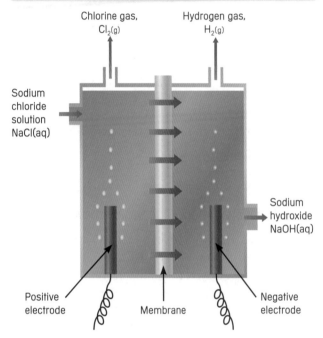

HT Electrolysis Equations

Reactions that occur at the electrodes during electrolysis can be represented by **half-equations**.

For example, in the electrolysis of copper...
- copper is deposited at the **negative electrode**

$$Cu^{2+} + 2e^- \longrightarrow Cu_{(s)}$$

- chlorine gas is given off at the positive electrode. (Remember that chlorine exists as molecules.)

$$2Cl^- \longrightarrow Cl_{2(g)} + 2e^-$$

N.B. When writing equations, remember to include the state symbols.

Quick Test

1. During electrolysis, which electrode do the positive ions move towards?
2. Fill in the missing words:
   a) When an ionic substance is melted or _____ in water, the _____ are free to move about.
   b) Passing an electric current through a molten ionic substance causes it to break down into _____.
   c) Electrolysis of sodium chloride solution produces sodium hydroxide, _____ and _____.

C2 Exam Practice Questions

1 This question is about the bonding present in sodium chloride (NaCl) and methane (CH_4).

  a) Complete the diagrams below to show the electron configurations in an atom of sodium and chlorine. **(2 marks)**

  Sodium Atom 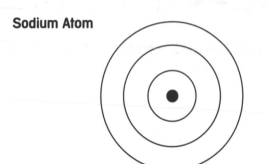   **Chlorine Atom**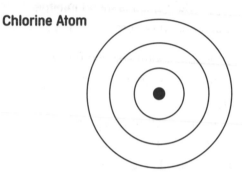

  b) When sodium and chlorine react an electron is transferred from the sodium atom to the chlorine atom. Explain why this occurs.

  .... **(1 mark)**

  c) What charge does the sodium have after it has transferred an electron to the chlorine atom?

  .... **(1 mark)**

  d) Draw a dot and cross diagram to represent the covalent bonding in a molecule of methane.

  (1 mark)

  e) Methane has a simple covalent structure. Explain why simple covalent structures have relatively low melting and boiling points.

  .... **(1 mark)**

2 An investigation was carried out into the effect of changing concentration on the rate of reaction.

  The following apparatus was set up.

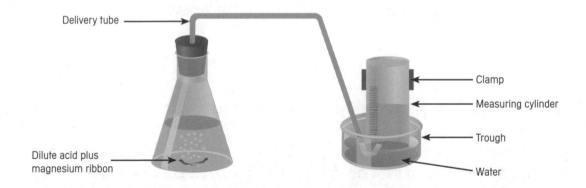

The concentration of acid was chosen. 5cm of magnesium ribbon was added and the time taken for $50cm^3$ of gas to be produced was recorded. A different concentration of acid was then used and the experiment repeated.

a) Why is it important that the mass of magnesium ribbon added is the same in each experiment?

.... **(1 mark)**

b) When the concentration of acid is increased how will the time taken to collect $50cm^3$ of gas change?

.... **(1 mark)**

c) Explain, in terms of particles, why a higher concentration of acid gives a faster rate of reaction.

....

.... **(1 mark)**

d) Name one other way of increasing the rate of reaction of magnesium with acid.

.... **(1 mark)**

HT **3** Nitric acid (HNO_3) is used to make fertilisers and explosives. Nitric acid reacts with ammonia (NH_3) according to the equation below:

$$HNO_3 + NH_3 \longrightarrow NH_4NO_3$$

a) Calculate the relative formula mass (M_r) of nitric acid.

.... **(1 mark)**

b) Calculate the mass of ammonium nitrate that would be expected to be formed if 6.3g of nitric acid reacts with an excess of ammonia.

....

....

.... **(1 mark)**

c) A student carried out the above experiment but only obtained 3g of ammonium nitrate. Calculate the percentage yield for this reaction. (If you were unable to do part **b)** of this question then use an answer of 5g to help you with part **c)**.

....

....

.... **(1 mark)**

d) Calculate the percentage by mass of nitrogen in ammonium nitrate.

....

....

.... **(1 mark)**

P2 Forces and their Effects

Resultant Forces

Forces are pushes or pulls. They are measured in units of force called a **newton** (N). Forces may…

- vary in **size**
- act in different **directions**.

Whenever two objects interact, the forces they exert on each other are **equal** and **opposite**. For example, when a stationary object rests on a surface it exerts a **downward force** on the surface due to the attractive force of gravity, which we call its **weight**. The surface it rests on exerts an equal and opposite **upward force**, called a **reaction force**.

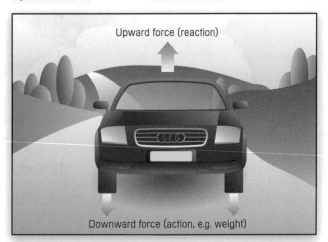

Upward force (reaction)

Downward force (action, e.g. weight)

If a number of different forces act on an object, these forces can be replaced by a single force. This single force has the same effect on the object as the original forces all acting together. This single force is called the **resultant force**.

A resultant force acting on an object may cause a change in its state of rest or motion.

If the resultant force acting on a **stationary** object is…

- zero, the object will remain **stationary**
- not-zero, the object will **accelerate** in the direction of the resultant force.

If the resultant force acting on a **moving** object is…

- zero, the object will continue to move at the **same speed** and in the **same direction**
- not-zero, the body will **accelerate** in the direction of the resultant force.

Forces and Motion

A resultant force acting on an object may cause a change in its state of rest or motion. The change of state of the object will depend on…

- the **size** of the resultant force (the bigger the resultant force the greater the acceleration)
- the **mass** of the object (the bigger the mass the smaller the acceleration).

The relationship between force, mass and acceleration is given by the following formula:

$$F = m \times a \quad \text{OR} \quad a = \frac{F}{m}$$

where F is the resultant force in newtons (N)
m is the mass in kilograms (kg)
a is the acceleration in metres per second squared (m/s^2)

This formula says that a force of 1 newton is needed to give a mass of 1kg an acceleration of 1m/s^2.

For example, a toy car of mass 800g accelerates with a force of 0.4N. Its acceleration is then $\frac{0.4N}{0.8kg}$ = 0.5 m/s^2 (remember, the mass must be in kg).

Speed

The **speed** of an object is just a measure of how fast it is moving. The speed of an object can be worked out if you know...

- the **distance** it travels
- the **time** taken to travel this distance.

You can calculate speed using the following formula:

$$s = \frac{d}{t}$$

where s is the speed in m/s
d is the distance travelled in metres
t is the time taken in seconds

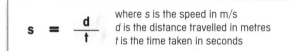
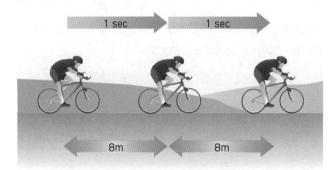

Speed can also be determined from the **slope** or **gradient** of a **distance–time graph**. The steeper the slope the greater the speed.

The graph shows:

1 A stationary person.

2 A person moving at a constant speed of 2m/s.

3 A person moving at a greater constant speed of 3m/s.

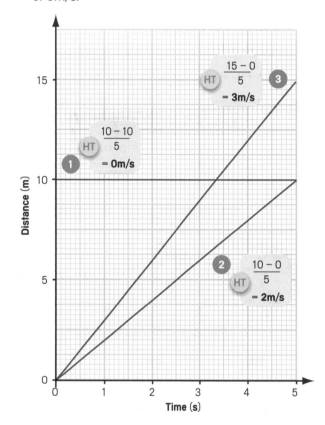

1 HT $\dfrac{10-10}{5}$ = 0m/s

2 HT $\dfrac{10-0}{5}$ = 2m/s

3 HT $\dfrac{15-0}{5}$ = 3m/s

Velocity

The **velocity** of an object and the speed of an object are **not** the same thing. The velocity of an object is its **speed in a certain direction**.

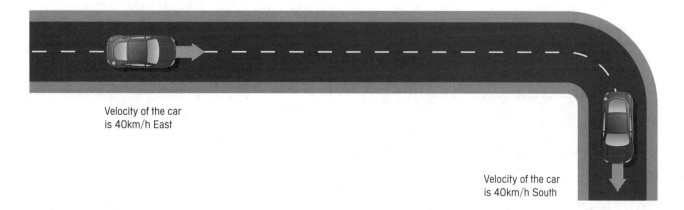

Velocity of the car is 40km/h East

Velocity of the car is 40km/h South

P2 Forces and their Effects

Acceleration

The **acceleration** of an object is the rate at which its **velocity changes**. It is a measure of how quickly an object speeds up or slows down.

To work out the acceleration of any moving object, you need to know…

*   the **change in velocity**
*   the **time taken** for this change to take place.

The acceleration of an object is given by:

$$\text{Acceleration} = \frac{\text{Change in velocity}}{\text{Time}}$$

The equation is as follows:

$$a = \frac{v - u}{t}$$

where a is the acceleration in m/s^2
v is the final velocity in m/s
u is the initial velocity in m/s
t is the time taken in seconds

The cyclist below increases his velocity by 2m/s every second. So, his acceleration is 2m/s^2.

*   His **velocity increases** by the same amount every second.
*   The actual **distance** travelled each second **increases**.

Deceleration is a negative acceleration. It describes an object that is slowing down. It is calculated using the same equation as for acceleration.

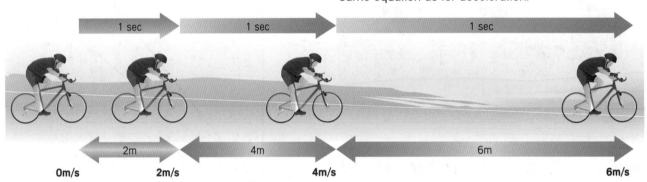

1 sec   1 sec   1 sec

2m   4m   6m

0m/s   2m/s   4m/s   6m/s

Velocity–Time Graphs

The velocity of an object can be represented by a **velocity–time graph**. The slope or gradient of a velocity–time graph gives the **acceleration** of an object.

The steeper the slope, the greater the acceleration.

(HT) The **area** underneath the line in a velocity–time graph represents the **total distance travelled**.

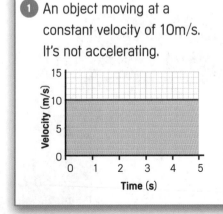

1 An object moving at a constant velocity of 10m/s. It's not accelerating.

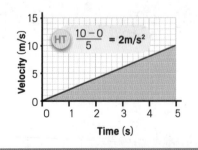

2 An object moving at a constant acceleration of 2m/s^2.

(HT) $\frac{10 - 0}{5} = 2$m/s^2

3 An object moving at a constant acceleration of -3m/s^2 (i.e. decelerating).

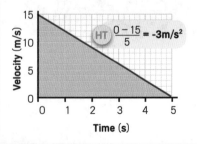

(HT) $\frac{0 - 15}{5} = -3$m/s^2

Forces and Braking

Friction is a force that occurs when…
- an object moves through a medium, e.g. air or water
- surfaces slide past each other.

Friction works against the object in the opposite direction to which it is moving, i.e. it's a **resistive force**.

When a vehicle travels at a steady speed the resistive forces (mainly air resistance) balance the **driving force**. The resultant force is the difference between the driving and resistive forces.

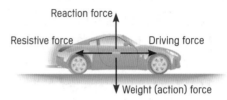

To increase a vehicle's top speed you need either a greater driving force or a reduction in the resistive force, e.g. by altering its shape (become more streamlined).

The **greater the speed** of the vehicle the greater the **braking force needed** to stop it in a certain time or certain distance.

Stopping Distance

The **stopping distance** of a vehicle depends on…
- the **thinking distance** (the distance travelled during the driver's reaction time)
- the **braking distance** (the distance travelled under the braking force).

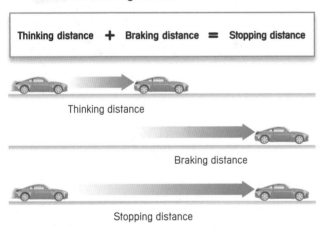

Thinking distance **+** Braking distance **=** Stopping distance

Thinking distance

Braking distance

Stopping distance

The overall stopping distance is increased if…
- the vehicle is travelling at **greater speeds**
- there are **adverse weather conditions**, e.g. wet roads, icy roads, poor visibility, etc
- the **driver is tired** or under the influence of **drugs** or **alcohol** or is **distracted** (e.g. mobile phone) and can't react as quickly as normal
- the **vehicle is in poor condition**, e.g. under-inflated tyres, worn brakes.

Friction forces between the **brakes** and the **wheel**, and between the **wheel** and the **road surface** reduce the kinetic energy of the vehicle. This **kinetic energy** is transformed into heating the brakes resulting in an **increase** in **brake temperature**. If a vehicle's wheels lock when braking, a skid results. Overheating can result in brake failure.

Forces and Weight

All falling objects experience two forces:
- A **downward force**, called **weight** (W).
- An **upward frictional force**, e.g. air resistance or drag through a fluid (R).

Although weight always remains the same, the **faster** an object moves through the air or fluid the **greater** the frictional force that acts on it.

The weight of an object is the force exerted on its mass by **gravity** (sometimes called **gravitational field strength**). Weight is measured in newtons.

To calculate the weight of an object the following equation is used:

$$W = m \times g$$

where the weight W is in newtons
m is the mass in kg and g is the gravitational field strength in newtons per kg (N/kg) (N.B this g has the same units as acceleration, i.e. m/s^2)

Terminal Velocity

An object falling through the air or a fluid will initially accelerate because of the force due to gravity. Eventually the resultant force will be zero as the **weight** and **resistive forces balance**. At this point the object will move at a steady speed, called its terminal velocity.

If a skydiver jumps out of an aeroplane, the speed of their descent can be considered in two separate parts:

* **Before** the parachute opens
* **After** the parachute opens.

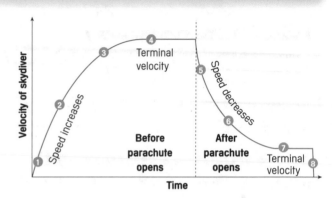

Before the Parachute Opens:

1. The skydiver accelerates due to the force of gravity.
2. The skydiver experiences frictional force due to air resistance in the opposite direction. At this point W is greater than R, so the skydiver continues to accelerate.
3. Speed increases, so does R.
4. R increases until it is the same as W. At this point the **resultant force is zero**. There is no more acceleration and the skydiver falls at a constant speed called the **terminal velocity**.

After the Parachute Opens:

5. The resistive force R is now greatly increased and is far bigger than W.
6. The increase in R decreases the skydiver's speed. As speed is reduced so is the value of R.
7. R decreases until it is the same as W. The forces balance for a second time and the skydiver falls at a steady speed although slower than before. This is a **new terminal velocity**.

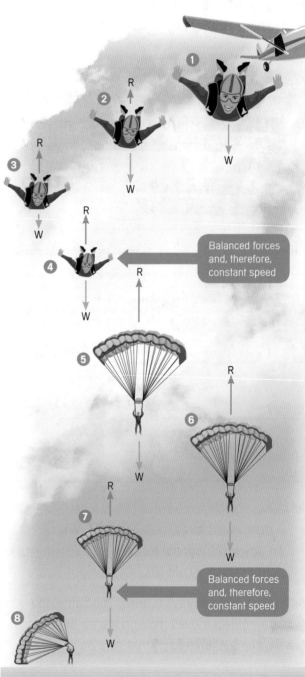

Forces and Elasticity

A force acting on an object may cause the object to change its shape.

A force applied to an object that's able to recover its original shape when the force is removed is said to be **elastic**, e.g. a spring.

When a force is applied to a spring, work is done in stretching the spring. The energy stored is called **elastic potential energy**. When the force is removed, the energy stored is used to bring the object back to its original shape.

For elastic objects, like springs, the extension is directly proportional to the force applied, provided that the limit of proportionality is not exceeded. The equation for this is as follows:

$$F = k \times e$$

where F is the applied force in newtons
e is the extension in metres
k is the proportionality constant, called the **spring constant**, measured in units of N/m

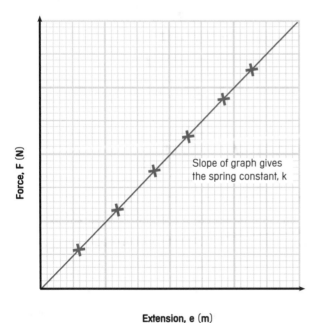

Slope of graph gives the spring constant, k

Force, F (N)

Extension, e (m)

Beyond the elastic limit the object permanently deforms.

Quick Test

1. If there is a not-zero resultant force acting on a toy car, what effect will be observed?
2. What additional property does velocity have over speed?
3. How is a constant deceleration shown on a velocity–time graph?
4. What two forces balance when a car travels at a constant speed?
5. What two factors affect a car's overall stopping distance?
6. What is meant by 'terminal velocity'?
7. Write an equation to express the balanced forces when terminal velocity is reached.
8. What is the name given to energy that is stored in an elastic spring?

P2 The Kinetic Energy of Objects

Forces and Energy

When a force causes an object to move through a distance, work is done on the object and energy is transferred. Both **work done** (W) and energy transferred (E) are measured in joules (J).

The amount of work done, force and distance are related by the equation:

$$W = F \times d$$

where W is the work done in joules, F is the force applied in newtons and d is the distance moved in the direction of the force in metres

Work done against frictional force mainly transfers energy to the surroundings and heats them.

Power is the work done in a given time and is given by the equation:

$$P = \frac{E}{t}$$

where P is the power in watts, E is the energy transferred in joules and t is the time taken in seconds

Gravitational Potential Energy

Gravitational potential energy is the energy that an object has due to its vertical position in the gravitational field. Work is done against the gravitational force and the object gains gravitational potential energy resulting from an increase in its height.

Gravitational potential energy can be calculated using the equation:

$$E_p = m \times g \times h$$

where E_p is the change in gravitational potential energy (joules), m is the mass in kilograms, g is the gravitational field strength in N/kg (approx. 10 on Earth), and h is the change in height in metres

Kinetic Energy and Momentum

The **kinetic energy** of an object is the energy it has due to its motion. Kinetic energy depends on…

- the **mass** of the object
- the **speed** of the object.

Kinetic energy can be calculated using the equation:

$$E_k = \frac{1}{2} \times m \times v^2$$

where E_k is the kinetic energy in joules, m is the mass in kilograms and v is the speed in m/s

A moving car has kinetic energy because it has both mass and speed. If it moves at greater speed it has more kinetic energy despite the mass being the same. A lorry moving at the same speed as a car will have greater kinetic energy due to its much greater mass.

During collisions the kinetic energy can be dissipated by using energy-absorbing devices such as air bags, crumple zones, seat belts and side impact bars in cars.

Unlike friction brakes, regenerative braking transfers unused energy back into useful electrical energy to recharge batteries and increase the overall efficiency of a vehicle, e.g. in hybrid cars and electric trains.

Momentum is a fundamental property of moving objects. It depends on…

- the **mass** of the object
- the **velocity** of the object.

Momentum can be calculated using the formula:

$$p = m \times v$$

where p is the momentum in kilograms metre per second (kg m/s), m is the mass in kilograms and v is the velocity in m/s

A moving car has momentum as it has both mass and velocity (speed in a certain direction). If the car moves with greater velocity, then it has more momentum providing its mass is the same.

For example, a car of mass 1200kg is moving with a velocity of 20m/s. Its momentum is 1200kg x 20m/s = 24000kg m/s. If the car moves with a new velocity of 30m/s then its new momentum is 1200kg x 30m/s = 36000kg m/s.

Key Words   *Work done • Gravitational potential energy • Kinetic energy • Momentum*

Conservation of Momentum

Momentum (like velocity) has…

- **size** (magnitude)
- **direction**.

The direction of movement is important when undertaking calculations involving momentum.

For example:

- Car A of mass 1400kg (moving from left to right) has a velocity of 20m/s to the right and, consequently, a momentum of 28000kg m/s to the right.
- Car B of mass 1400kg (moving from right to left) has a velocity of 20m/s to the left, i.e. -20m/s and momentum of 28000kg m/s to the left or -28000kg m/s with respect to car A because it is moving in the opposite direction to car A. Its momentum is -28000kg m/s.

A fundamental principle of momentum is that in a closed system, i.e. where no other external forces act, the total momentum before an event is equal to the total momentum **after the event**. This is called the **conservation of momentum**.

Example

Two cars are travelling in the same direction along a road. Car A collides with the back of car B and they stick together. Calculate their velocity after the collision.

Before

After

Momentum before collision:

= Momentum of A + Momentum of B

= (mass x velocity of A) + (mass x velocity of B)

= (1200kg x 20m/s) + (1000kg x 9m/s)

= 24 000kg m/s + 9000kg m/s

= 33 000kg m/s

Momentum after collision:

= Momentum of A and B

= (mass of A + mass of B) x (velocity of A + B)

= (1200 + 1000) x v

= 2200v

Since momentum is conserved:

Total momentum before = Total momentum after

$$33\,000 = 2200v$$

$$\text{So, } v = \frac{33\,000}{2200}$$

$$= \textbf{15m/s}$$

Quick Test

1. What is 'work done' in terms of energy?
2. What is the name given to the amount of work done per second?
3. Calculate the kinetic energy of a car of mass 1000kg moving at 12m/s.
4. What is meant by the phrase 'conservation of momentum'?

P2 Currents in Electrical Circuits

Static Electricity

Some insulating materials can become electrically charged when they are rubbed against each other. Unless it is **discharged**, the electrical charge, called static electricity, stays on the material.

Static electricity builds up when electrons (negative charge) are 'rubbed off' one material on to another. The material…

- **gaining** electrons becomes **negatively** charged
- **losing** electrons becomes **positively** charged.

For example, a Perspex rod rubbed with a cloth becomes positively charged and an ebonite rod rubbed with fur becomes negatively charged.

Perspex Rod   **Ebonite Rod**
Rubbed with Cloth   **Rubbed with Fur**

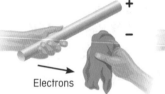

When two electrically charged objects are brought together they exert a force on each other. They are either attracted or repelled:

- Materials with the **same** charge **repel** each other, e.g. Perspex and Perspex.
- Materials with **different** charges **attract** each other, e.g. Perspex and ebonite.

Perspex Rod repels   **Perspex Rod attracts**
a Perspex Rod   **an Ebonite Rod**

Although static is transferred between the surfaces of materials, electrical charges can also readily move through some materials, e.g. metals. This is because there are electrons from their atoms that are free to move through the metal structure (called **free electrons**).

Current

An electric current through a circuit is a **flow of electric charge**. The **size** of the electric current is the **rate of flow** of electric charge and is given by the following equation:

$$I = \frac{Q}{t}$$

where *I* is the current in amperes (amps, A), *Q* is the charge in coulombs (C) and *t* is the time in seconds (s)

Potential Difference

An electric current will flow through an electrical component (or device) if there is a potential difference (voltage) across the ends of the component. The potential difference between two points in an electric circuit is the work done (energy transferred) per coulomb of charge that passes between the points.

Potential difference is given by the equation:

$$V = \frac{W}{Q}$$

where *V* is the potential difference in volts (V), *W* is the work done in joules (J) and *Q* is the charge in coulombs (C)

Resistance

The amount of current that flows through a component depends on...
- the **potential difference** across the component
- the **resistance** of the component.

All components resist the flow of current through them. Resistance is a measure of how hard it is to get a current through a component at a particular potential difference. Resistance is measured in **ohms**, which have the symbol Ω.

The greater the resistance of the components...
- the **smaller** the **current** that flows for a particular potential difference

OR

- the **greater** the **potential difference** needed to maintain a particular current.

To calculate the current, potential difference or resistance the following equation is used:

$$V = I \times R$$

where V is the potential difference in volts (V), I is the current in amperes (amps A), and R is the resistance in ohms (Ω)

Circuits

In a circuit...
- the potential difference (p.d) provided by cells connected in series is the sum of the potential difference of each cell
- the potential difference is measured in volts (V) using a **voltmeter** connected in **parallel**
- the current is measured in amperes (A) using an **ammeter** connected in **series**.

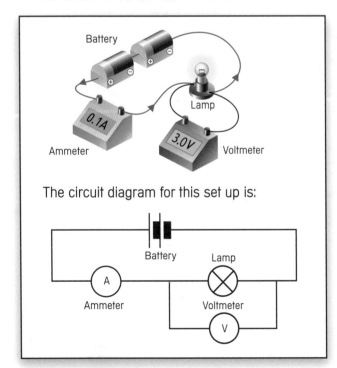

Battery

0.1A

Ammeter

Lamp

3.0V

Voltmeter

The circuit diagram for this set up is:

Battery

Lamp

A

Ammeter

Voltmeter

V

Standard symbols are used to represent the various components in circuit diagrams. For example:

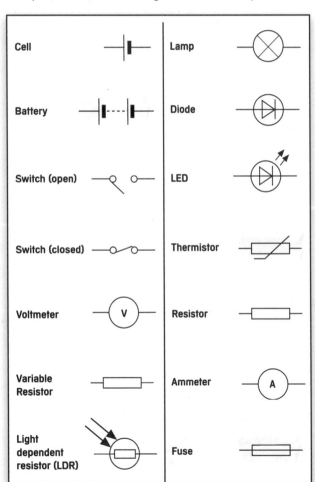

Cell		Lamp	
Battery		Diode	
Switch (open)		LED	
Switch (closed)		Thermistor	
Voltmeter		Resistor	
Variable Resistor		Ammeter	
Light dependent resistor (LDR)		Fuse	

Current–Potential Difference Graphs

Current–potential difference graphs show how the current through a component varies with the potential difference across it.

The resistance of a **light dependent resistor** (**LDR**) depends on the amount of light falling on it. Its **resistance decreases** as the amount of **light** falling on it **increases**. This allows more current to flow.	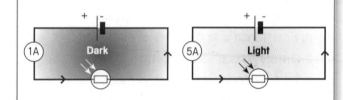
The resistance of a **thermistor** depends on its **temperature**. Its **resistance decreases** as the **temperature** of the thermistor **increases**. This allows more current to flow.	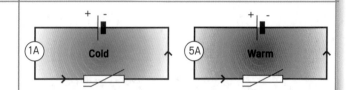
As long as the temperature of the **resistor** stays constant, the current through the resistor is directly proportional to the potential difference across the resistor. This is regardless of which direction the current is flowing, i.e. if one doubles, the other also doubles.	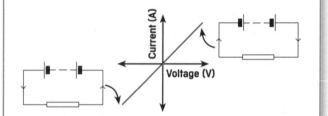
As the temperature of the **filament lamp** increases, and the bulb gets brighter, then the resistance of the lamp increases. (HT) This is due to the greater vibrations of the metallic ions in the filament wire gradually preventing the flow of free electrons.	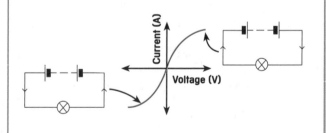
A **diode** allows a current to flow through it in **one direction only**. It has a very high resistance in the reverse direction so no current flows. A light emitting diode (LED) emits light when a current flows through it in the forward direction. There is an increasing use of LEDs for lighting as they use a much smaller current than other forms of lighting and are most cost-effective.	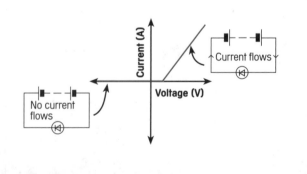

Currents in Electrical Circuits P2

Series Circuits

For components connected **in series**:

- The total resistance (R) is the sum of the resistance of each component, $R = R_1 + R_2$
- There is the same current through each component, $I = I_1 = I_2$
- The total potential difference of the supply (V) from the battery is shared between the components, $V = V_1 + V_2$

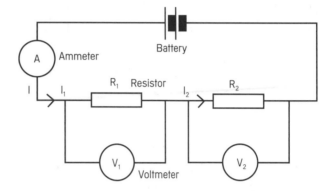

Parallel Circuits

For components connected **in parallel**:

- The potential difference across each component is the same, $V = V_1 = V_2$
- The total current through the whole circuit is the sum of the currents through the separate components, $I = I_1 + I_2$

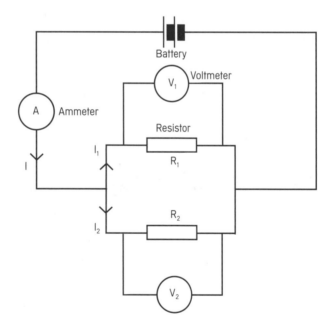

Quick Test

1. What will two materials with the same charge do if they are brought together?
2. What name is given to the rate of flow of electric charge?
3. What unit is used for resistance?
4. If identical components are connected in parallel, what is the current through each component?
5. What happens to the resistance of a thermistor when its temperature increases?

P2 Mains Electricity

Direct and Alternating Currents

Wet cell, dry-cell non-rechargeable, and dry-cell rechargeable batteries supply a current that always passes in the same direction. This is called **direct current (d.c)**. The trace for d.c. on a cathode ray oscilloscope is a straight line.

Alternating current (a.c.) is one that is constantly changing direction (oscillates). The trace for a.c. on a cathode ray oscilloscope is a wave. The **period** and **amplitude** of the wave form determines the nature of the a.c. supply.

Mains electricity is an a.c. supply. In the UK it has a frequency of 50 cycles per second (50 hertz) and is about 230V. The voltage, if it isn't used safely, can kill.

Example

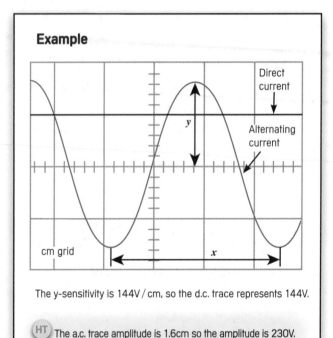

The y-sensitivity is 144V/cm, so the d.c. trace represents 144V.

HT The a.c. trace amplitude is 1.6cm so the amplitude is 230V. The trace period is 3.2cm and the x sensitivity is 6.25ms/cm, so the time period is 20ms. This gives a frequency of 1/T or 50Hz

The Three-pin Plug

Most electrical devices are connected to the mains electricity supply using a **cable** and a **three-pin plug**. The plug is inserted into a socket on the ring main circuit.

The materials used for the plug and cable are designed to reduce risk of electrocution. The following are the main properties of the cable and plug:

- The **inner cores** of the wires are made of copper because it's an excellent conductor.
- The **outer layers** of the wires are made from flexible **plastic** because it's a **good insulator**.
- Wires comprise of either **two-core** or **three-core** cable. The three-core carries an earth wire.
- The **pins** of the plug are made from **brass** because it's a **good conductor**, strong and stiff.
- The casing is made from plastic or rubber because both are good insulators.

The Three-pin Plug

Fuse

Casing

Earth wire (green and yellow)

Neutral wire (blue) – carries current away from appliance

Cable grip – secures cable in the plug

5A

Live wire (brown) – carries current to the appliance

Cable

Direct current (d.c.) • Alternating current (a.c.)

Circuit Breakers and Fuses

If an electrical fault occurs there is an increase in the current flow. A **fuse** or **circuit breaker** in the circuit provides a disconnection in the live wire, effectively switching off the circuit.

Depending on the type of electrical appliance, the plug will be fitted with fuses that have **different ratings**, e.g. 3A, 5A, 13A. When the current in the **fuse** wire exceeds the rating of the fuse it will **melt, breaking the circuit**.

The thicker the cable, the **higher** the rating of the **fuse value**. Fuses have to be replaced each time the circuit is overloaded.

Some modern circuits are protected by using circuit breakers, which **automatically** break an electric circuit if it becomes overloaded. Circuit breakers are easily **reset** by pressing a button.

Some circuits are protected by **Residual Current Circuit Breakers** (**RCCBs**). These operate by detecting a **difference** in the current between the **live** and **neutral** wires. These devices operate much faster than a fuse.

Earthing

Devices that have outer **metal cases** are usually **earthed**. The outer case of an electrical appliance is connected to the earth pin in the plug through the earth wire.

The earth wire and fuse work together to protect the appliance (and the user).

If a fault occurs:
1. The case will become live.
2. The current will then 'flow to earth' through the earth wire as this offers least resistance.
3. This overload of current will cause the fuse to melt (or circuit breaker to trip), breaking the circuit.
4. The appliance (and user) are therefore protected.

Some appliances, e.g. drills, are **double insulated**, and therefore have no earth wire connection.

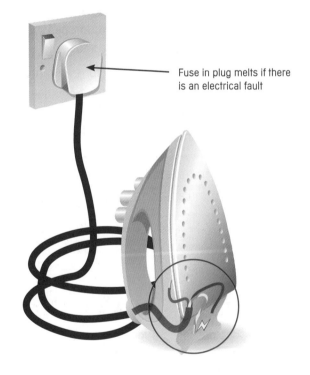

Fuse in plug melts if there is an electrical fault

Energy Transfer

When an electrical **charge** (current) flows through a **resistor** (e.g. electrical device or appliance), the resistor gets **hot**. Some of the electrical energy is used but a lot of energy is wasted, which usually heats the surroundings.

In a filament bulb only 5% of the energy goes into light, the remaining 95% is wasted as heat energy. Less energy is wasted in power-saving lamps such as **Compact Fluorescent Lamps** (CFLs).

Key Words   Fuse • Circuit breaker • Residual Current Circuit Breaker (RCCB) • Double insulated

P2 Mains Electricity

Power

The rate at which energy is transferred by an appliance is called the **power**.

Power can be calculated using the following formula:

$$P = \frac{E}{t}$$

where *P* is the power in watts (W)
E is the energy transferred in joules (J)
t is the time in seconds (s)

A much more useful expression for power is one that connects power with the current and potential difference:

$$P = I \times V$$

where *P* is the power in watts (W)
I is the current in amperes (amps, A)
V is the potential difference in volts (V)

For example, a vacuum cleaner rated at 1100W and using mains electricity (230V), provides a current of $\frac{P}{V} = \frac{1100W}{230V} = 4.8A$. A fuse with a rating of 5A would be suitable for the safe operation of this appliance.

Important factors to consider when buying household appliances (e.g. fridges, washing machines and dishwashers) are their energy efficiency and their power rating. Equally important is their ease of maintenance and location, which helps reduce heat loss and maintain efficiency.

Charge

The amount of electrical **charge** that passes any point in a circuit is measured in **coulombs** (C). As charge passes through a device energy is transferred. The amount of energy transferred for every coulomb of charge depends on the size of the potential difference. The **greater** the potential difference, the **more** energy is transferred per coulomb.

(HT) The energy transferred can be calculated using the following formula:

$$E = V \times Q$$

where *E* is energy in joules (J), *V* is the potential difference in volts (V) and *Q* is the charge in coulombs (C)

Example

If a circuit has a potential difference of 1.5V, and a charge of 24C passes through it, how much energy is transferred?

Energy transferred = p.d. x Charge
= 1.5V x 24C
= **36 joules**

Remember, the charge gained this energy from the battery. It was transferred to the bulb whilst the circuit was switched on

Quick Test

1. Name two devices that give protection when an electrical fault occurs.
2. What colour is the earth wire in a three-pin plug?
3. A washing machine has a power rating of 2300W and uses mains electricity. What is the size of fuse that must be used in the three-pin plug?
4. A filament lamp transfers 690 joules of energy using mains electricity. What is the amount of charge that passes through the circuit?

Atoms

Atoms are the basic particles from which all matter is made. The basic structure of an **atom** is an extremely tiny central nucleus composed of **protons** (positive charge) and **neutrons** (no electrical charge) surrounded by **electrons** (negative charge).

This model of the atom, the nuclear model, was based on results from the Rutherford and Marsden scattering experiments and replaced the earlier 'plum pudding model'.

A Fluorine Atom

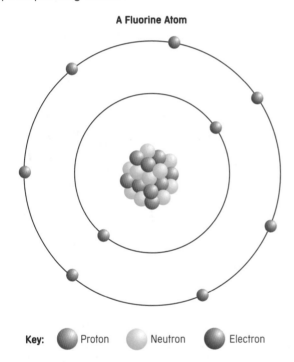

| Key: | ● Proton | ○ Neutron | ● Electron |

In an atom the number of **electrons** is equal to the number of **protons** in the nucleus. So the atom as a whole has no electrical charge and is therefore electrically **neutral**. In the nuclear model most of the atom is empty space.

Atoms of different elements have different numbers of protons (and electrons).

The number of protons in an atom therefore defines the element.
- The number of **protons** in an atom is called its **atomic number**.
- The number of **protons and neutrons** in an atom is called its **mass number**.

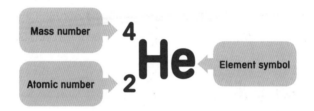

Atomic Particle	Relative Mass	Relative Charge
Proton	1	+1
Neutron	1	0
Electron	$\frac{1}{2000}$	-1

The size of the nucleus is about 10^5 times smaller than the size of the atom.

Isotopes and Ions

Some atoms of the **same element** can have different numbers of **neutrons**. These are called **isotopes**.

For example, oxygen has three common isotopes, $^{16}_{8}O$, $^{17}_{8}O$ and $^{18}_{8}O$ with only $^{16}_{8}O$ being stable.

Atoms may also lose or gain electrons to form charged particles called **ions**. An atom that has gained electrons is called a **negative ion**. An atom that has lost electrons is a **positive ion**. Positively charged ions attract negatively charged ions and can form a strong bond, e.g. sodium chloride (salt).

Unstable Nuclei

Isotopes of atoms that have too many or too few neutrons form **unstable nuclei**. The nuclei may disintegrate by **randomly emitting radiation**. Atoms of these isotopes are radioactive (also called **radioactive** isotopes, radioisotopes or radionuclides). The process of disintegration is called **radioactive decay**.

Radioactive decay is all around us and is commonly referred to as background radiation. Background radiation is **not harmful** to our health as it occurs in very small amounts or radiation doses. Actual levels depend on where you live and what occupation you do.

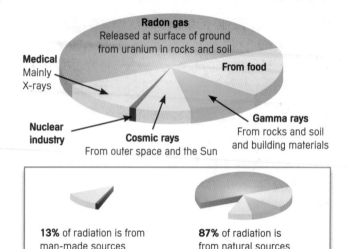

Radon gas
Released at surface of ground from uranium in rocks and soil

Medical
Mainly X-rays

Nuclear industry

Cosmic rays
From outer space and the Sun

From food

Gamma rays
From rocks and soil and building materials

13% of radiation is from man-made sources

87% of radiation is from natural sources

Radioactive Decay

The radioactive decay process can result in the formation of a different atom with a different number of protons.

Three examples of this are...
- **alpha radiation (α)**
- **beta radiation (β)**
- **gamma (γ) radiation**.

Unlike alpha and beta decay, gamma emissions have no effect on the internal structure of the nucleus. Gamma radiation is a form of electromagnetic radiation that carries away any surplus energy from the nucleus.

Gamma radiation can be used for example in radiotherapy to kill cancer cells and shrink malignant tumours.

Radiation can cause damage to living cells. Inside the body alpha radiation is most dangerous as it is easily absorbed by cells, whereas beta and gamma radiation are less harmful as they easily pass through the cells. Outside the body the roles are reversed, with alpha radiation less harmful and beta and gamma radiation considerably more dangerous.

An alpha particle is a huge particle. It's identical to a **helium nucleus**, consisting of **two protons** and **two neutrons**, and symbolised by $^4_2$**He**. In alpha decay a completely new atom is formed.

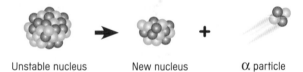

Unstable nucleus   New nucleus   α particle

(HT) For example, Radium-226 decays by alpha emission to form Radon-222, a radioactive gas. The nuclear equation for this decay process is:

$$^{226}_{88}\text{Ra} \longrightarrow {}^{222}_{86}\text{Rn} + {}^{4}_{2}\text{He}$$

Beta Decay

In beta decay the original atom decays by changing a **neutron** into a **proton** and an **electron**.

The newly formed high-energy electron is ejected from the nucleus. To distinguish it from orbiting electrons around an atom, the electron emitted is called a beta particle with the symbol β.

Alpha Decay

In alpha decay the original atom decays by ejecting an **alpha particle** from its nucleus.

Unstable nucleus   New nucleus   β particle

Beta Decay (Cont.)

(HT) For example, Radon-222 also decays by alpha emission to give Polonium-218. This new atom is also radioactive and decays by beta emission to give Astatine-218. The nuclear equation for this decay is:

$$^{218}_{84}\text{Po} \longrightarrow \ ^{218}_{85}\text{At} + \ ^{0}_{-1}\beta$$

Notice how the top and bottom numbers balance on either side of the equation. The beta particle carries a negative charge.

Beta decay is used in medical imaging (PET scans) as tracers to highlight and diagnose cancers.

Ionisation and Penetration Power

When radiation collides with **neutral** atoms or molecules in a substance, the atoms or molecules may become charged due to electrons (the outer electrons surrounding the atoms or molecules) being 'knocked out' of the orbiting structure during the collision.

This alters their structure, leaving the atoms or molecules as **ions** (i.e. atoms with an electrical charge) or as **charged particles**.

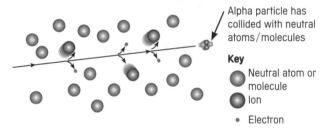

Alpha particle has collided with neutral atoms / molecules

Key
- Neutral atom or molecule
- Ion
- Electron

Each type of emitted radiation (alpha, beta, gamma) has a different…
- degree of **ionising power**
- ability to **penetrate** materials
- **range** in air
- amount of **deflection** in electric and magnetic fields.

(HT) The degree of deflection depends on…
- the **relative masses** of the alpha particle compared to the beta particle
- the **charge** on each particle (+2 for alpha particle and -1 for the beta particle).

Particle	Description	Ionising Power	Penetration	Affected by Electric and Magnetic Fields
Alpha (α)	• Helium nucleus • Positive particle	Strong	Stopped by paper or skin or 6cm of air	Yes, but opposite to beta particles
Beta (β)	• Negative electron	Weak	Stopped by 3mm of aluminium	Yes, bent strongly, but opposite to alpha particles
Gamma (γ)	• Electromagnetic radiation • Very short wavelength	Very weak	Reduced but not stopped by lead	No

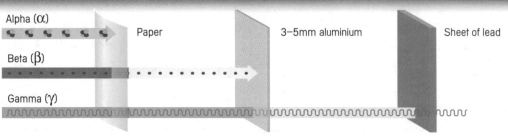

Alpha (α)   Paper   3–5mm aluminium   Sheet of lead

Beta (β)

Gamma (γ)

Half-life

The **half-life** of a radioactive isotope is a measurement of the time it takes for the rate of decay (count-rate) to halve **or** the time required for half of the original population of radioactive atoms to decay.

A radioactive isotope that has a very long half-life remains active for a very long time.

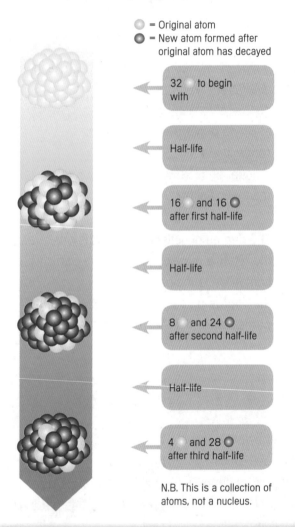

= Original atom
= New atom formed after original atom has decayed

32 ● to begin with

Half-life

16 ● and 16 ○ after first half-life

Half-life

8 ● and 24 ○ after second half-life

Half-life

4 ● and 28 ○ after third half-life

N.B. This is a collection of atoms, not a nucleus.

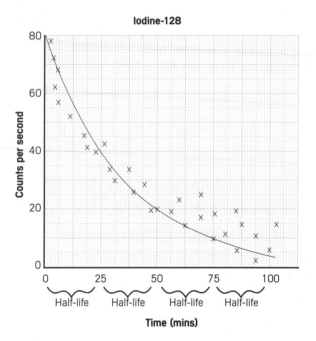

Iodine-128

Counts per second

Time (mins)

Half-life   Half-life   Half-life   Half-life

The graph shows the count rate (using a Geiger counter) from a sample of radioactive Iodine-128 against time. It shows that...

- the initial count rate was 80 counts per second
- after 25 minutes (1 half-life) the count rate has fallen to 40 counts per second
- after two half-lives the count rate is only 20 counts per second.

Radioactive isotopes can have half-lives between fractions of a second to tens of thousands of years.

The choice of which radioisotope to use depends on its purpose and whether it is used internally or externally to diagnose or to treat.

Quick Test

1. An element, E, is represented in the following way: $^{M}_{A}E$. What do the letters M and A stand for?
2. Radioactive nuclei can decay by emitting alpha particles. What is an alpha particle?
3. In a magnetic field alpha particles, beta particles and gamma radiation behave differently. Which particle is not deflected at all? Give your reasons.
4. What is meant by the term 'half-life' of a radioactive isotope?

Nuclear Fission

Nuclear fission is the **splitting** of an atomic nucleus. It's used in nuclear reactors to release **energy** to make electricity.

The two fissionable substances in common use in nuclear reactors are **uranium-235** and **plutonium-239**.

For fission to occur the uranium-235 or plutonium-239 nucleus must first **absorb** a neutron. The nucleus then becomes unstable and splits into two smaller nuclei, releasing two or three more neutrons and a lot of energy.

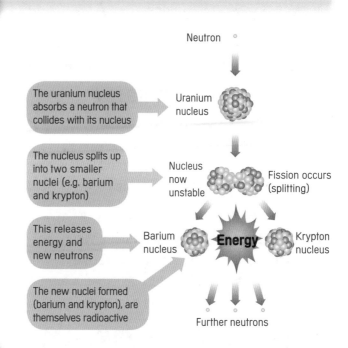

Neutron

The uranium nucleus absorbs a neutron that collides with its nucleus

Uranium nucleus

The nucleus splits up into two smaller nuclei (e.g. barium and krypton)

Nucleus now unstable

Fission occurs (splitting)

This releases energy and new neutrons

Barium nucleus

Energy

Krypton nucleus

The new nuclei formed (barium and krypton), are themselves radioactive

Further neutrons

Chain Reaction

The neutrons released in fission may themselves go on to collide with other uranium-235 or plutonium-239 nuclei, producing further neutrons and energy in a process called a **chain reaction**. Nuclear reactors control the rate of this chain reaction to release the energy required.

The new neutrons produced by nuclear fission can each cause a new fission. This is a **chain reaction**. It carries on and on and on

Energy

Energy   **Energy**   **Energy**

The energy is released and heats the surroundings. Each fission reaction only releases a tiny amount of energy, but there are billions and billions of reactions every second

P2 Nuclear Fission and Nuclear Fusion

Nuclear Fusion

Nuclear fusion is the **joining** together of two or more atomic nuclei to form a larger atomic nucleus.

To achieve nuclear fusion a lot of energy is required.

A nuclear fusion reaction (when started) will release more energy than it uses. This makes it **self-sustaining**, i.e. some of the energy produced is used to drive further fusion reactions.

An example is the fusion of two isotopes of hydrogen, called **deuterium** and **tritium**. When they are forced together under high pressure, the deuterium and tritium nuclei fuse together to form a **helium atom** and a **neutron** and a lot of energy.

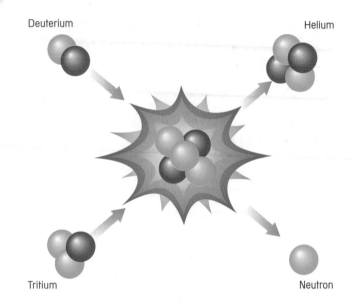

Deuterium

Helium

Tritium

Neutron

Star Formation

Nuclear fusion is the process by which energy is released in stars. In the core of our nearest star, the Sun, hydrogen is being continuously converted into helium through nuclear fusion. This process provides the energy to keep the Sun hot and to allow life on Earth.

Stars, like our Sun, form when enough dust and gas from space are pulled together by gravitational forces, which always attract each other. This forms a **nebula** where a **protostar** is then formed.

Forcing material together increases the temperature and density, and nuclear fusion reactions start releasing huge amounts of energy. Eventually the attractive **gravitational forces** balance with the **repulsive forces** produced by radiation to make a star **stable**.

The newly formed star becomes a **main sequence** star. It will remain like this for many millions or billions of years until its supply of hydrogen runs out.

During the formation process smaller masses within the protostar may be attracted by the dominant larger mass to become **planets**.

Star Formation

During its time on the main sequence, a star will produce all of the naturally occurring elements through the fusion process up to iron.

Nuclear fusion • Nebula • Protostar • Main sequence

Stellar Evolution

Eventually the hydrogen within a star runs out. What happens next is determined by the size (mass) of the star.

Stars about the Size of the Sun

1. Star leaves the main sequence and becomes a **red giant**.
2. It continues to cool before collapsing under its own gravity to become a **white dwarf**.
3. It continues to cool and loses its brightness to become a **black dwarf**.

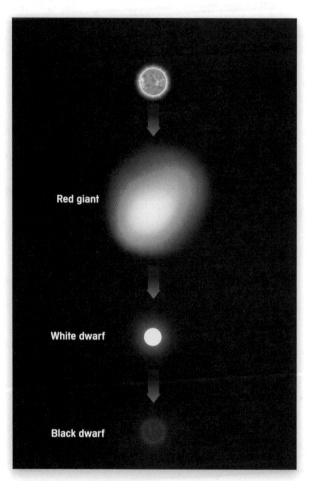

Stars Much Bigger than the Sun

1. Star leaves the main sequence and become a **red super giant**.
2. It cools but shrinks very rapidly and explodes as a **supernova**. This explosion releases massive amounts of energy, dust and gas into space, and forms elements heavier than iron.
3. Depending on the precise mass of the remnants either a **neutron star** or a **black hole** is formed.
4. The dust and gas form new stars.

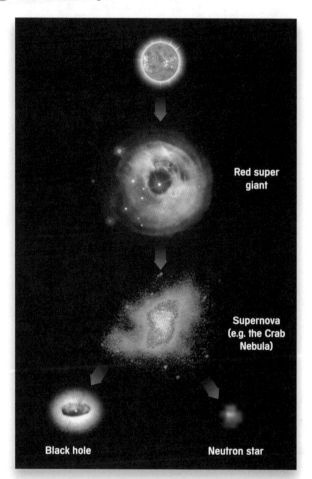

Quick Test

1. Give an example of a fissionable material used in nuclear power stations.
2. What is meant by the term 'chain reaction'?
3. Where is nuclear fusion the dominant energy releasing process?
4. What is the heaviest element that can be produced in Sun-type stars by nuclear fusion?

---

1 **a)** The diagram shows the position of the Sun in the main sequence of events when the Sun's energy begins to run out. The size of the circles provide an indication as to the size of star involved.

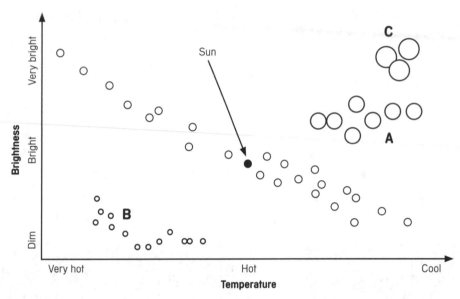

i) What type of star does the Sun become, shown at position A?

.... **(1 mark)**

ii) After millions of years this star becomes similar to those shown at position B. What type of star is this?

.... **(1 mark)**

(iii) Using the graph, describe the two key features of each type of star.

Star A : ....

Star B : .... **(2 marks)**

(iv) What is the name of the star at the end of the Sun's life cycle?

.... **(1 mark)**

(b) A star that is much larger than our Sun takes a very different path. Part of its history takes it to position C shown in the diagram.

(i) What is the name for this type of star? .... **(1 mark)**

(ii) What is the name given to the catastrophic event that occurs eventually in these stars?

.... **(1 mark)**

(iii) What are the two possible final outcomes from such events?

....

.... **(2 marks)**

2 A skydiver jumps from an aeroplane and free falls without opening their parachute.

X

Weight

a) In the diagram of the free falling skydiver what is the name of the force X?

.... **(1 mark)**

b) Explain what happens to the force X as the skydiver accelerates.

.... **(1 mark)**

c) If the mass of the skydiver is 60kg, calculate their change in gravitational potential energy between leaving the aeroplane and falling 250m (take g = 10m/s^2).

....

.... **(2 marks)**

d) In reality, not all the gravitational energy is converted into kinetic energy, however assuming that it is, calculate the speed of the skydiver after 250m of free fall.

....

....

.... **(3 marks)**

e) After 250m of free fall the skydiver no longer accelerates or speeds up. What do we call this point?

.... **(1 mark)**

f) What value does the force X now have at (and beyond) this point when the skydiver is still in free fall?

.... **(1 mark)**

Answers

Biology

Page 13
1. Nucleus
2. In ribosomes
3. In chloroplast
4. Enzymes
5. Diffusion
6. **Accept one from:** Carbon dioxide; Waste products

Page 15
1. Tissue
2. An organ
3. The leaf
4. The nervous system
5. So they can carry out different jobs

Page 18
1. Chlorophyll
2. From the air
3. Nitrate ions
4. Temperature, carbon dioxide concentration and light intensity
5. Random sampling with a quadrat and sampling along a transect

Page 21
1. Protein
2. Temperature and pH
3. Stomach, pancreas and small intestine
4. Sugar or glucose
5. Lipase
6. The liver. Its function is to neutralise acid (and emulsify fats)
7. Protease and lipase

Page 23
1. Mitochondria
2. Aerobic respiration
3. It is breathed out / exhaled
4. Lactic acid
5. Oxygen debt

Page 25
1. Mitosis
2. XY
3. Meiosis
4. 23

Page 29
1. Alleles
2. In human embryos and adult bone marrow
3. Recessive
4. XX
5. A small section of DNA that controls a characteristic.
6. **Accept any suitable answer, e.g.:** Some animals were soft-bodied and did not leave traces behind; Fossils lost due to geological activity (weathering and erosion); Conditions for fossilisation rare – predators eat body / body rots, etc.
7. Heterozygous

Exam Practice Answers
Pages 30-31
1. a) Chloroplasts, cell wall and vacuole (**all 3 correct for 1 mark**)
   b) Energy
2. a) water; oxygen
   b) In chloroplast
   c) **Accept two from:** Produce fat or oil and stored; Produce cellulose to strengthen cell wall; Produce proteins
3. a) Quadrat
   b) **Accept three from:** Set an area; Place quadrats randomly within the area; Count the number of different plant species; Repeat several times
   c) Transect
4. The movement of particles from a region of high concentration to a region of low concentration.
5.

Enzyme	Where is it Produced?	What Does it Digest?	What Does it Produce?
Amylase	Salivary glands, pancreas and small intestine	Starch	**Sugars**
Protease	**Stomach**, pancreas and small intestine	**Proteins**	Amino acids
Lipase	Pancreas and small intestine	Lipids (fats and oils)	Glycerol and **fatty acids**

6. **Accept three from:** Heart rate increases; Breathing rate increases; Arteries to muscles dilate; Blood flow to muscles increases; Supply of oxygen and glucose increases as well as removal of carbon dioxide
7. 46; 23
8. **Accept two from:** Changes to the environment; New predators; New diseases; New, more successful competitors; Single catastrophic event
9. a) Anaerobic respiration
   b) Oxygen debt. Lactic acid is oxidised to carbon dioxide and water

Chemistry

Quick Test Answers
Page 33
1. Two or more atoms chemically combined together
2. Covalent
3. **a)** melting points / boiling points; intermolecular
    b) electricity; move
Page 37
1. The layers of atoms are able to slide over each other.
2. Thermo-setting polymers consist of cross-links between the polymer chains. Thermo-softening polymers do not.
3. **Any two from:** In computers; Catalysts; Coatings; Highly selective sensors; Stronger and lighter construction materials; New cosmetics, e.g. suntan creams and deodorants
Page 39
1. Mass number
2. Isotopes
Page 41
1. **a)** 22%
    b) 35%
    c) 21%
2. **a)** 2
    b) 0.19
3. **a)** 56g
    b) 414g
Page 44
1. **Any two from:** Instrumental methods are accurate; Sensitive and rapid; Useful when dealing with small quantities
2. **Any two from:** The reaction may be reversible; Some of the product may be lost when it's separated from the reaction mixture; Some of the reactants may react in ways different to the expected reaction.
3. The relative molecular mass of a compound
4. 11g
Page 48
1. Particles are given more energy so there will be more collisions and more energetic collisions (i.e. more collisions where the particles collide with sufficient energy to react).
2. increase; reduce / decrease
3. The surroundings
4. It will be endothermic
Page 51
1. Precipitation reactions
2. **a)** Alkalis
    b) Fertilisers
    c) Water

Page 53
1. Negative (cathode)
2. **a)** dissolved; ions
    b) elements
    c) hydrogen; chlorine

Exam Practice Answers
Pages 54-55
1. **a)** Sodium Atom   Chlorine Atom

    b) So that both atoms end up with a full outer shell of electrons, which makes them stable
    c) +1
    d)

    H
    H   C   H
    H

    e) There are weak forces between the molecules
2. **a)** So that only one variable is changed (in this case concentration)
    b) The time taken will decrease
    c) At a higher concentration there will be more particles and so there will be a greater likelihood of particles colliding causing a reaction to take place.
    d) **Accept one from:** Increasing the temperature; Increasing the surface area of the magnesium, e.g. using a powder; Adding a catalyst
3. **a)** 63
    b) 8g
    c) $\frac{3}{8}$ x 100 = 37.5% (If you used 5g, then $\frac{5}{8}$ x 100 = 62.5%)
    d) $\frac{28}{80}$ x 100 = 35%

Answers

Quick Test Answers

Page 61
1. An acceleration in the direction of the resultant force
2. Direction
3. Negative gradient or slope
4. Friction force and driving force
5. Reaction time (thinking distance) and braking distance
6. Constant speed reached when the upward resistive force balances the downward force (weight)
7. $R = W$ or $R = mg$ at terminal velocity
8. Elastic potential energy

Page 63
1. Energy transferred
2. Power
3. 72000J
4. Total momentum before = Total momentum after

Page 67
1. Repel
2. Current
3. Ohms or Ω
4. It is shared equally between the components
5. Decreases

Page 70
1. Fuse; RCCB
2. Green and yellow
3. 13A fuse
4. 3 coulombs

Page 74
1. M is the mass number – the number of protons and neutrons; A represents the atomic number – the number of protons.
2. A helium nucleus or 2 protons and 2 neutrons
3. Gamma radiation; Carries no charge and is not deflected by a magnetic field
4. The time taken for half of the radioactive nuclei to decay into stable nuclei

Page 77
1. Uranium-235 or plutonium-239
2. Neutrons released from the initial reaction go on to interact with other nuclei producing even more neutrons each time.
3. Within stars
4. Iron

Exam Practice Answers

Pages 78-79
1. a) i) Red giant
   ii) White dwarf
   iii) Star A: brighter, cooler; Star B: dimmer, hotter
   iv) Black dwarf
   (b) i) Red super giant
   ii) Supernova explosions
   iii) Neutron star; Black hole
2. a) Air resistance / drag
   b) X becomes larger
   c) $E_p = mgh = 60 \times 10 \times 250; = 150000J$
   d) $E_k = \frac{1}{2}mv^2 = 150000J$; $v^2 = 2 \times \frac{150000}{60} = 5000$; so that $v = \sqrt{5000} \approx 71m/s$
   e) Terminal velocity
   f) 600N

Glossary

Biology

Aerobic respiration – respiration that uses oxygen

Allele – an alternative form of a particular gene

Amino acids – the smaller soluble sub-units that join to form proteins

Amylase – an enzyme that breaks down starch

Anaerobic respiration – respiration that takes place without oxygen

Bile – a greenish-yellow fluid produced by the liver

Catalyst – a substance that increases the rate of a chemical reaction without being changed itself

Cell – a fundamental unit of a living organism

Chlorophyll – the green pigment found in most plants; responsible for photosynthesis

Chloroplast – a tiny structure in the cytoplasm of plant cells that contains chlorophyll

Chromosome – long molecule found in the nucleus of all cells; made from DNA

Cystic fibrosis – genetic disorder caused by recessive alleles affecting cell membranes

Cytoplasm – the substance found in living cells (outside the nucleus) where chemical reactions take place

Denatured – where the special shape of an enzyme is changed (by excessive temperature or pH) so it no longer functions

Differentiation – to make / become different

Diffusion – the mixing of two substances through the natural movement of their particles from a high concentration to a low concentration

Dilate – to widen or enlarge

DNA (deoxyribo nucleic acid) – nuclei acid molecules that contain genetic information and make up chromosomes

Dominant (allele) – an allele that only needs to be present once in order to be expressed; represented by a capital letter

Enzyme – a protein that speeds up a reaction (a biological catalyst)

Epidermis – outer layer, e.g. skin

Extinction – where all individuals of a species have died out

Fatigued – extremely tired due to excessive activity on muscles carrying out anaerobic respiration and the build up of lactic acid

Fertilisation – the fusion of the male and female gametes

Fossil – the remains of animals / plants preserved in rock

Gamete – a specialised sex cell formed by meiosis

Gene – part of a chromosome made up of DNA; controls a certain characteristic

Glycogen – a form of carbohydrate in which sugars are stored in the body for energy

Lactic acid – a compound produced when cells respire without oxygen (i.e. anaerobically)

Lipase – an enzyme that breaks down fat into fatty acids and glycerol

Meiosis – cell division that forms daughter cells with half the number of chromosomes of the parent cell

Mendel – an Austrian monk who studied inheritance

Mitochondria – the structure in the cytoplasm where energy is produced from chemical reactions

Mitosis – cell division that forms two daughter cells, each with the same number of chromosomes as the parent cell

Multicellular – an organism consisting of many cells

Nucleus – the control centre of the cell

Organ – a group of tissues working together

Phloem – tissue for transporting sugars around a plant

Photosynthesis – the chemical process that uses light energy to produce glucose in green plants

Polydactyly – genetic disorder caused by a dominant allele where people have extra fingers or toes

Protease – an enzyme used to break down proteins into amino acids

Quadrat – a square frame (usually between $0.25m^2$ and $1m^2$) used for sampling organisms in their natural environment (usually plants, aquatic organisms, e.g. barnacles, and some slowly mobile animals such as insects)

Quantitative – data involving quantities or measurements, e.g. distribution of organisms

Recessive (allele) – an allele that will only be expressed if there are two present; represented by a lower case letter

Glossary

Respiration – the process of converting glucose into energy inside cells

Ribosome – small structure found in the cytoplasm of living cells where protein synthesis takes place

Sexual reproduction – when new individuals are produced that aren't genetically identical to the parents; involving the fusion of gametes

Specialised – adapted for a particular purpose

Stem cell – a human embryo cell or adult bone marrow cell that has yet to differentiate

Tissue – a group of cells that have a similar structure and function

Transect – a fixed line along which sampling of populations, such as species abundance, are measured

Vacuole – a fluid-filled sac found in cytoplasm

Xylem – tissue for transporting water and minerals in plants

Yeast – a single-celled fungus; a microorganism

Zygote – a cell formed by the fusion of the nuclei of a male sex cell and a female sex cell (gametes)

HT **Genotype** – the combination of alleles an individual has for a particular gene, e.g. BB, Bb or bb

Heterozygous – an individual who carries two different copies of the allele, e.g. Bb

Homozygous – an individual who carries two copies of the same allele, e.g. BB or bb

Oxygen debt – oxygen deficiency caused by anaerobic respiration during intense / vigorous exercise

Phenotype – the expression of the genotype, i.e. the characteristic shown

Speciation – where populations have become so different that successful interbreeding is longer possible

Chemistry

Acid – a compound that has a pH lower than 7

Activation energy – the minimum amount of energy required to cause a reaction

Alkali – a compound that has a pH higher than 7

Atomic number – the number of protons in an atom

Catalyst – a substance that increases the rate of a chemical reaction without being changed itself

Chromatography – a technique used to separate different compounds in a mixture according to how well they dissolve a particular solvent

Compound – a substance consisting of two or more elements chemically combined together

Covalent bond – a bond between two atoms, in which both atoms share one or more electrons

Cryolite – an aluminium-containing ionic compound; used in the electrolytic recovery of aluminium

Current – the flow of electric charge through a conductor

Electrode – a piece of metal or carbon that allows electric current to enter and leave during electrolysis

Electrolysis – the process by which an electric current causes a solution to undergo chemical decomposition

Electrolyte – the molten or aqueous solution of an ionic compound used in electrolysis

Electron – a negatively charged subatomic particle that orbits the nucleus

Element – a substance that consists of only one type of atom

Endothermic – a reaction that takes in heat from its surroundings

Exothermic – a reaction that gives out heat to its surroundings

Insoluble – a substance that will not dissolve in a solvent

Ion – a charged particle formed when an atom gains or loses electrons

Ionic bond – the bond formed between two (or more) atoms when one loses, and another gains, electrons to become charged ions

Ionic compound – a compound formed when two (or more) elements bond ionically

Isotopes – atoms of the same element that contain different numbers of neutrons

Mass number – the total number of protons and neutrons present in an atom

Mole (mol) – the molar mass of a substance, i.e. the mass in grams of 6×10^{23} particles

Monomer – a small hydrocarbon molecule containing a double bond

Nanoscience – dealing with materials that have a very small grain size, in the order of 1–100nm

Neutralisation – a reaction between an acid and a base that forms a neutral solution (i.e. pH 7)

Neutron – a subatomic particle found in the nucleus of an atom that has no charge

Oxidation – a reaction involving the gain of oxygen, the loss of hydrogen, or the loss of electrons

pH – a measure of acidity or alkalinity

Precipitate – an insoluble solid formed in a precipitation reaction

Precipitation – a type of reaction in which a solid is made when two liquids are mixed

Proton – a positively charged subatomic particle found in the nucleus

Reduction – a reaction involving the loss of oxygen, the gain of hydrogen, or the gain of electrons

Relative atomic mass (A_r) – the average mass of an atom of an element compared with a twelfth of the mass of a carbon atom.

Relative formula mass (M_r) – the sum of the atomic masses of all atoms in a molecule

Reversible reaction – a reaction in which products can react to re-form the original reactants

Salt – the product of a chemical reaction between a base and an acid

Smart alloy – an alloy that can change shape and then return to its original shape

Soluble – a substance that can dissolve in a solvent

Thermo-setting (polymer) – polymer chains that are joined together by cross-links

Thermo-softening (polymer) – polymer chains that are tangled together

Yield – the amount of a product obtained from a reaction

Physics

Acceleration – the rate of change of velocity; units of m/s^2; $a = \frac{F}{m}$

Alpha particle – a helium nucleus; 2 protons and 2 neutrons

Alternating current (a.c.) – an electric current that reverses its direction of flow repeatedly

Atom – the smallest part of an element; the building blocks of matter

Atomic number – the number of protons in an atom

Background radiation – radiation that is around us; predominantly from natural sources

Beta particle – an electron emitted from the nucleus

Black dwarf – the end point in a star's life cycle for stars similar in mass to the Sun

Black hole – the remains of a star after a supernova explosion

Chain reaction – a reaction that is self-sustaining

Charge – a property of elementary particles; comes in two forms – positive and negative

Circuit breaker – a safety device that automatically breaks an electric circuit when it becomes overloaded

Conservation of momentum – the total momentum before and after a collision is the same

Current–potential difference graph – graph used to show how the current through a component varies with the potential difference across it

Current (electric) – the rate of flow of electric charge; units of amperes (amps) A; $I = \frac{Q}{t}$

Diode – an electrical device that allows electric current to flow in one direction

Direct current (d.c) – an electric current that flows in one direction

Distance–time graph – a graph showing the distance travelled against time taken; the gradient of the line represents speed

Double insulated – electrical appliances that need no earth connection

Elastic – a force applied to an object that recovers its original shape when the force is removed

Elastic potential energy – the energy stored in a stretched spring

Electron – a fundamental particle with a charge of -1; orbits the nucleus

Force – a push or pull acting on an object

Fuse – a thin wire that overheats and melts when overloaded to break an electric circuit; comes with different ratings

Glossary

Friction – a resistive force acting between two surfaces

Gamma radiation – high energy, high frequency electromagnetic radiation

Gravitational field strength – the value of 'g' at a particular height above the Earth's surface

Gravitational potential energy – the energy that an object has due to its position in a gravitational field; $E_p = m \times g \times h$; units of joule (J)

Half-life – the average time taken for half of the unstable atoms in a radioactive material to decay

Ion – an atom that has gained or lost an electron

Ionising power – the ability of particles or electromagnetic radiation to ionise other neutral atoms or molecules

Isotopes – atoms of the same element that contain a different number of neutrons, but the same number of protons

Kinetic energy – the energy possessed by an object due to its motion: $E_k = \frac{1}{2} \times m \times v^2$; units of joule (J)

Main sequence – the position of a star that determines its life cycle; based on a star's mass

Mass number – the number of protons and neutrons in an atomic nucleus

Momentum – a fundamental quantity that is a measure of the state of motion of an object; product of mass and velocity; $p = m \times v$; units of kg m/s

Nebula – a general term for a 'fuzzy' patch of sky composed of interstellar dust and gas

Neutron – a subatomic particle found in the nucleus; has no charge

Neutron star – the remains of a star after a supernova explosion

Nuclear fission – the splitting of an atomic nucleus into two smaller nuclei with the emission of neutrons and energy

Nuclear fusion – the joining together of light atomic nuclei with the emission of energy; main energy process in stars

Potential difference – the energy transfer by unit charge passing from one point to another; measured in volts (V)

Power – work done or energy transferred in a given time; units of watts (W); $P = \frac{E}{t}$ and $P = I \times V$

Proton – a subatomic particle found in the nucleus; has a charge of +1

Protostar – the initial gas and dust that forms a star

Radioactive – the decay of unstable nuclei by the emission of alpha or beta particles or gamma radiation

Red giant – part of a star's life cycle after leaving the main sequence for stars with masses similar to our Sun's mass

Red super giant – part of a star's life cycle after leaving the main sequence for stars with masses much larger than our Sun

Residual Current Circuit Breaker (RCCB) – automatic device for breaking a circuit; based on detecting a difference in current between live and neutral wires

Resistance – opposition to the flow of electric current; units of ohms

Resistor – an electrical device that resists the flow of an electric current

Resultant force – the combined effect of all forces acting on an object

Speed – the rate at which an object moves

Spring constant – the gradient of the line of a graph of force against extension; units of N/m

Static electricity – the loss or gain of electrons on the surface of a material

Stopping distance – the sum of the thinking distance and braking distance

Supernova – catastrophic explosion of a red super giant

Terminal velocity – the constant maximum velocity reached by a falling object; weight balances the frictional forces

Thermistor – a resistor whose resistance varies with temperature

Unstable nuclei – nuclei that are radioactive and emit radiation

Velocity – the speed of an object in a specified direction

Velocity–time graph – a graph showing velocity against time; the gradient of the line represents acceleration; the area under the line represents distance (displacement)

Weight – the gravitational force exerted on an object with mass; $W = m \times g$

White dwarf – part of a star's life cycle after its red giant stage

Work done – the product of force and distance moved; $W = F \times d$

Electronic Structure of the First 20 Elements

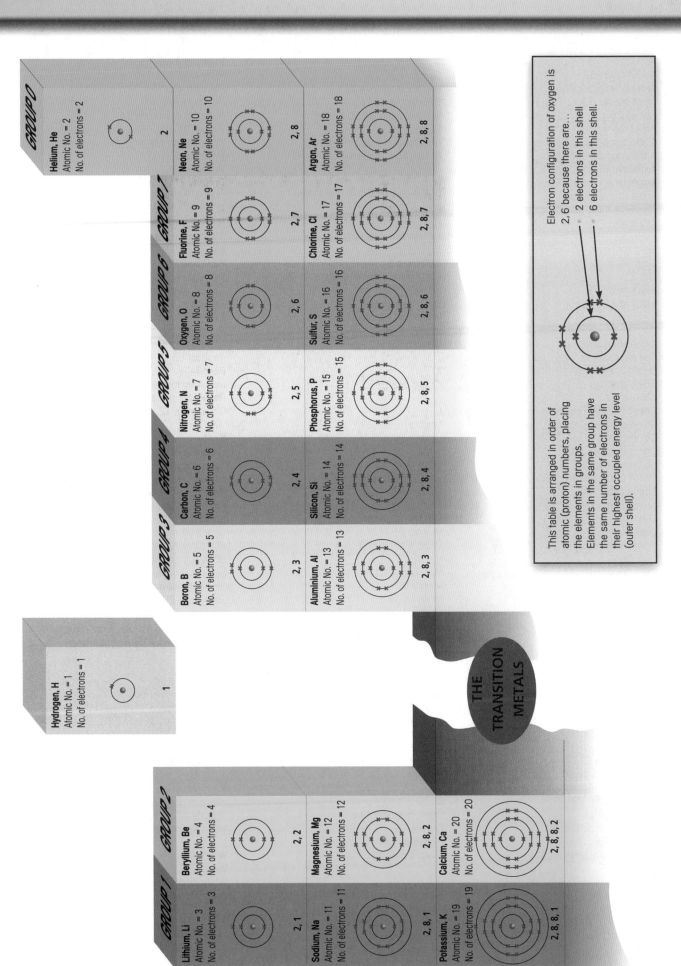

GROUP 0

Helium, He
Atomic No. = 2
No. of electrons = 2

2

Neon, Ne
Atomic No. = 10
No. of electrons = 10

2, 8

Argon, Ar
Atomic No. = 18
No. of electrons = 18

2, 8, 8

GROUP 7

Fluorine, F
Atomic No. = 9
No. of electrons = 9

2, 7

Chlorine, Cl
Atomic No. = 17
No. of electrons = 17

2, 8, 7

GROUP 6

Oxygen, O
Atomic No. = 8
No. of electrons = 8

2, 6

Sulfur, S
Atomic No. = 16
No. of electrons = 16

2, 8, 6

GROUP 5

Nitrogen, N
Atomic No. = 7
No. of electrons = 7

2, 5

Phosphorus, P
Atomic No. = 15
No. of electrons = 15

2, 8, 5

GROUP 4

Carbon, C
Atomic No. = 6
No. of electrons = 6

2, 4

Silicon, Si
Atomic No. = 14
No. of electrons = 14

2, 8, 4

GROUP 3

Boron, B
Atomic No. = 5
No. of electrons = 5

2, 3

Aluminium, Al
Atomic No. = 13
No. of electrons = 13

2, 8, 3

Electron configuration of oxygen is
2, 6 because there are...
• 2 electrons in this shell
• 6 electrons in this shell.

This table is arranged in order of atomic (proton) numbers, placing the elements in groups.
Elements in the same group have the same number of electrons in their highest occupied energy level (outer shell).

Hydrogen, H
Atomic No. = 1
No. of electrons = 1

1

THE TRANSITION METALS

GROUP 1

Lithium, Li
Atomic No. = 3
No. of electrons = 3

2, 1

Sodium, Na
Atomic No. = 11
No. of electrons = 11

2, 8, 1

Potassium, K
Atomic No. = 19
No. of electrons = 19

2, 8, 8, 1

GROUP 2

Beryllium, Be
Atomic No. = 4
No. of electrons = 4

2, 2

Magnesium, Mg
Atomic No. = 12
No. of electrons = 12

2, 8, 2

Calcium, Ca
Atomic No. = 20
No. of electrons = 20

2, 8, 8, 2

Index